Plain Talk

About

Drinking Water

Sixth Edition

Plain Talk

About

Drinking

Water

Sixth Edition

Answers to Your Questions About the Water You Drink

DR. JAMES M. SYMONS
NANCY E. McTIGUE

American Water Works Association

Plain Talk About Drinking Water, Sixth Edition
Copyright © 2025 American Water Works Association
All rights reserved.

Disclaimer

American Water Works Association
6666 West Quincy Avenue
Denver, CO 80235-3098
303.794.7711

Sr. Manager —Product Acquisition & Development: Geoffrey S. Shideler
Director of Publishing: John Fedor
Technical Editor: Wiley
Technical Editing Review: Suzanne Snyder
Design and Production: Gillian Wink
Illustrations: Melanie Yamamoto
Cover Images: Leigh Prather and Lanka 69/Shutterstock.com

Library of Congress Cataloging-in-Publication Data

Names: McTigue, Nancy E. author | Symons, James M. author
Title: Plain talk about drinking water / by Nancy E. McTigue and James M.
   Symons.
Description: Sixth edition. | Denver, CO : American Water Works
   Association, [2025] | Includes index. | Summary: "This expanded and
   revised edition provides answers to more than 200 of the most commonly
   asked questions about drinking water and related issues,   including
   information on health, bottled water, reservoirs, conservation, testing,
   emerging contaminants, and more"-- Provided by publisher.
Identifiers: LCCN 2025030799 (print) | LCCN 00024296374 (ebook) | ISBN
   9781647172442 | ISBN 9781613007679 pdf
Subjects: LCSH: Drinking water--Miscellanea | Water quality--Miscellanea
Classification: LCC TD353 .S96 2025 (print) | LCC TD353 (ebook) | DDC
   363.6/1--dc23/eng/20250819
LC record available at https://lccn.loc.gov/2025030799
LC ebook record available at https://lccn.loc.gov/2025030800

ISBN 978-1-64717-244-2
ISBN, electronic 978-1-61300-767-9
DOI: https://doi.org/10.12999/AWWA.20244-6E

Foreword

Shortly after I started my work at AWWA, I realized that I had taken safe drinking water for granted most of my life. This is a common phenomenon I have witnessed for staff who come from other sectors. We have a resource that is critical to our health, our community, and our environment, but it is something that remains in the background of our busy lives.

Like so many others, I used to turn on my tap, send water down the drain, and pay my bill each month. I didn't think about my water utility at all—unless something wasn't right. But now that I work in the water sector, I understand how massively complex and technically challenging it is to produce and deliver safe, reliable water. I certainly have more appreciation for the water coming from my tap (and going down my drain), and I'm surrounded by subject matter experts who work daily to ensure we always have safe and reliable water service.

Potable water by its very nature is a technical subject. A simple question like "Is my water safe to drink?" can quickly lead to a very complicated answer with jargon, nuances, and technicalities. And while this book's original edition was published in 1992, we now have an endless internet, countless social media channels, and so much collective information that it can be overwhelming to process it all. So, if customers want answers to their questions about tap water, where do they even begin? This book, now in its sixth edition, provides simple and factual answers to common questions about drinking water.

AWWA recognizes the significance of public involvement and customer communication in the water sector, and our policy stance underscores the need for open, honest, and proactive dialogue with the communities we serve. This approach not only builds trust and cooperation; it also enhances the credibility and reputation of water utilities.

Whether you want to learn how to communicate technical water terms to people in your community, or you are a curious individual looking for answers to questions about your water, I hope this book helps you. And my other hope is that, if you're not in the water sector, you'll come to appreciate that members of the AWWA community are willing and able to discuss technical, complex subjects in plain terms that anyone can understand. Because even though drinking water is a technical subject, water is important *to* everyone, and therefore it's important to be able to communicate about it *with* everyone.

Geoffrey S. Shideler

Senior Manager | Product Acquisition & Development

American Water Works Association

March 2025

Table of Contents

Chapter 7: Regulations, Reporting, and Water Security

Preface

The first edition of this book was published more than 30 years ago. The goal of that book was to provide explanations for some of the questions that surround drinking water. Where does my tap water come from? How is it treated? Why does ice float? And to explain these things and more, in simple but factual language—using "plain talk."

Some things about drinking water haven't changed, but many things have changed since the first edition of this book was published in 1992. The US Environmental Protection Agency (USEPA) (established in 1972) was in its infancy in 1992. Since that time, the USEPA has continuously updated what is being regulated and how water utilities must operate. Bottled water wasn't the multibillion-dollar industry that it is today. Now, many people say they get most of their drinking water from plastic bottles. Furthermore, our ability to detect substances in water using scientific instruments has dramatically increased in the past 30 years. Not only have the kinds of contaminants we're concerned about changed, but also the amount of each contaminant that our water plants need to remove. Maybe the biggest difference is that when the first edition was published, the Internet was not yet available to the public. That didn't happen until 1993. Now, the Internet is readily available and provides all kinds of information at our fingertips.

Drinking water is still an amazing, fascinating, and vital substance, so this book has been continually revisited through many editions. Water must not only be available and of good quality day in and day out but literally all the time. This sixth edition builds on the work of the last five editions and seeks to answer questions about new and emerging contaminants and concerns while keeping the basic information that has been vetted through the years.

The nearly 250 questions and answers are organized by general topic area.

The first chapter, "Health," is the largest, filled with facts central to the question of "Is my water safe to drink?" Read about water's contribution to health, the occurrence and health effects of common chemical and microbial contaminants, treatment chemicals, travel precautions, and more. Other chapters provide information about

- taste, odor, and appearance—the aesthetics of your tap water;
- home facts—ways in which water is used in a home as well as information about bottled water, home treatment, and costs;
- sources—groundwater, surface water, reservoirs, pollution issues, and source water protection;
- distribution—pipes, hydrants, tanks, and leaks;
- conservation—the why, what, and how of saving water;
- regulations and testing—the role of the federal, state, and provincial government; and
- fascinating facts—beyond drinking water, from solid to liquid to gas.

Our knowledge and understanding of many of the topics covered here will continue to grow, regulations and testing requirements will continue to change, and new subjects will arise after the publication of this edition. Therefore, we have provided within the answers, and in the Acknowledgments section, the best resources on the Internet. These websites are regularly updated and will offer additional and up-to-date information.

We hope you enjoy learning some new things about drinking water!

Nancy E. McTigue

February 2025

Acknowledgments

Throughout all editions of this book, answers to the listed questions have been checked and updated by countless water professionals using the best resources. Internet sites of organizations of particular value to this book include

- AWWA (www.awwa.org);
- US Environmental Protection Agency (USEPA) (www.epa.gov);
- Centers for Disease Control and Prevention (CDC) (www.cdc.gov);
- Food and Drug Administration (FDA) (www.fda.org);
- NSF International (www.nsf.org);
- The Water Research Foundation (WRF) (www.waterrf.org);
- American National Standards Institute (ANSI) (www.ansi.org);
- Canadian Federal–Provincial–Territorial Committee on Drinking Water (CDW) (www.canada.ca); and
- Water Quality Association (WQA) (www.WQA.org).

Special thanks for the review of this edition are extended to Carrie Lewis, previous superintendent of the Milwaukee Water Works, Milwaukee, Wis., and general manager of the Portland (Maine) Water District. Carrie's expertise in drinking water quality science, water treatment, and water system financial and security issues and her management of two large municipal water systems provided invaluable insight.

Also, thanks to Dr. James M. Symons (retired), the Cullen Distinguished Professor Emeritus of Civil Engineering at the University of Houston and the original author and contributor to each of the subsequent editions, for continuing to provide his knowledge and expertise in answering the questions in this edition. His excitement about anything related to drinking water is obvious throughout the book.

Thanks are offered to Dr. David Cornwell, PE, president of Cornwell Engineering and adjunct professor in Environmental Engineering and Sciences at the University of Florida, who brought his extensive knowledge and understanding of water chemistry, environmental engineering, water treatment and regulations, and ice cubes to his review.

Thank you to Matthew Junker, public relations specialist at Municipal Authority of Westmoreland County (New Stanton, Pa.) and Monica Hoyt, client services manager at Carollo Engineers (Salt Lake City, Utah), both of AWWA's Public Affairs Council, who helped make the answers to the questions more focused on what consumers really want to know. And to Brent Alspach, director of applied research at Arcadis (Carlsbad, Calif.), who helped with answers to questions about the emerging issues in the field.

And finally, thank you to Madison Cornwell, who brought her unique perspective as a young professional to this project.

Chapter 1

Health

Water is life's matter and matrix, mother and medium. There is no life without water.

—Albert Szent-Gyorgyi

1. Is my tap water safe to drink?

Tap water from public water systems (PWSs) in the United States and Canada is among the safest in the world, and maintaining that quality is a priority for PWSs along with both the US Environmental Protection Agency (USEPA) and Health Canada. In the United States, 90% of the population gets its tap water from a PWS.

In the United States, the Safe Drinking Water Act (SDWA) sets the standards for drinking water quality and treatment; in Canada, the *Guidelines for Canadian Drinking Water Quality* (www.Canada.ca) provides the basis for drinking water quality parameters. In most cases, but not all, USEPA standards and Canadian guidelines for a particular contaminant are the same.

There are often reports in the news about contaminants found in public drinking water, and sometimes short-term disease outbreaks are traced to water supplies. But water utilities are required to monitor for and control more than 100 different parameters that may affect water at the tap—parameters that may negatively affect human health and that are likely to be present in

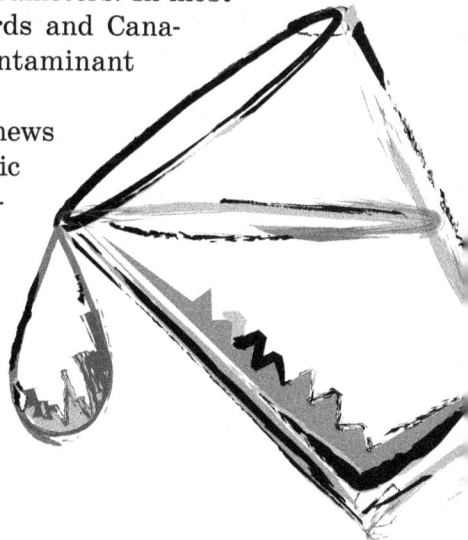

drinking water. If your tap water meets all the federal, state or provincial quality standards, it is considered safe to drink. Nearly all PWSs in the United States and Canada meet or exceed the standards set for these parameters.

As mentioned throughout this book, you should consult with your own doctor regarding any specific concerns.

2. How can I find out more about the quality of my tap water?

Sometimes it's hard to know if drinking water contaminants and problems discussed in the media affect your tap water. To find out, it's best to go to the source!

The USEPA requires all PWSs in the United States to issue a water quality report once or twice a year, depending on the size of the system. This document, the consumer confidence report (CCR), is sometimes called a water quality report. It is usually available on the utility website, which is a great place to get information on your own drinking water. You may have to call the utility to get a copy of the report if you can't find it on their website. It lists all the parameters your utility must monitor and compares your utility's results with the national standards. This report will tell you if your water utility meets all the requirements of the federal regulations. If your tap water meets all the federal quality standards, it is considered safe to drink. The report also tells you where your water comes from and how it is treated.

The USEPA's website, www.epa.gov, has a lot of information about these reports and can direct you to your utility's report. You can also call your utility to ask about a specific contaminant.

If you get your water from a well, and not from a PWS, you can contact your state or local health department, and they can direct you to laboratories that can test your well water.

3. Is potable water the same as safe water?

When it is considered safe to drink, water is called potable, which rhymes with "floatable." Water is considered safe to drink if it meets or exceeds all the federal, state, and provincial standards that are legally enforceable. In the United States, if your tap water does not meet any one of the standards, your water supplier must notify all its affected customers.

4. How much water should I drink to stay healthy?

Your body is about 60% water, which it uses to regulate its temperature, digest food, carry nutrients throughout the body, and flush waste out of the body. So, you need to be sure to drink enough water to keep the body hydrated.

Many healthy US residents rely on their thirst to guide them rather than following the old "eight to nine (8 ounce) glasses a day" rule, which is about 64 to 72 ounces (1.9 to 2.1 liters). But thirst alone should not be used as a guide for when to drink. By the time you become thirsty, it's possible to already be slightly dehydrated. Others would say the amount of your water intake should depend on your weight, where you live, and how active you are. Because 60% of your body is water, it is essential to replenish what water you lose through your breath, perspiration, urine, and bowel movements, which for most people is about 8 cups (1.9 liters) a day.

The Institute of Medicine of the National Academies (www.nam.edu) suggests that drinking 11 cups (2.6 liters) of water for women and 16 cups (3.8 liters) for men a day is essential. Some of that can come from juices, milk, soft drinks, and even solid foods. But this can vary based on factors such as your level of activity and the climate you live in.

Consumption of salty foods, diseases such as diabetes, and various medications can affect a person's thirst sensation. Finally, in proportion to body weight, babies need more fluids than adults. Consult with your doctor about the water needs of your baby.

5. Must all my water intake be plain water, or are drinks made with water OK?

Juice, milk, and soft drinks are almost all water, so they do count toward the required total daily fluid intake. Nutritionists often recommend tap water, however, because some other beverages contain chemicals, such as caffeine and alcohol, that act as a diuretic and do not help maintain fluid balance as well as other drinks. Vegetables and fruits also contain water that can help replenish your system and can account for about 20% of your fluid intake.

6. Does drinking water contain calories, fat, sugar, or cholesterol?

No. Water is the original "health" drink. You might have noticed that some bottled water has added nutrients such as vitamins. You should read the bottled water label carefully to see if those added nutrients include sugar, fructose, or dyes.

7. How long can a human go without water?

This depends on weather, shelter, altitude, activity level, and whether a person has access to food that contains water. For example, in hot conditions with no water, dehydration can set in within an hour and death can occur within two to three days. An infant left in a hot car or an adult exercising hard in hot weather can dehydrate, overheat, and die in just a few hours. In some conditions, a person can live up to 7–10 days without water, but water deprivation isn't something that should be tested! On the other hand, most adults can live three to six weeks without food, depending on how much fat is stored in their bodies at the onset of food deprivation. Water is essential to life!

4

8. Is it true that you can drink too much water?

Yes. Drinking too much water is uncommon, but water intoxication can occur and can even cause death. Too much water intake can result in the sodium in your blood being diluted, a condition called hyponatremia. This imbalance of electrolyte content makes the body's tissue cells swell and puts pressure on the brain and nerves, and ultimately, death can occur.

Endurance athletes, babies, and military trainees are potentially at risk from this condition.

9. Does anyone actually get sick from drinking tap water?

Yes. If a water system is compromised in some way, it can become contaminated by microbes or chemicals. Drinking water containing viruses, parasites, and bacteria can cause illness. *Giardia, Cryptosporidium,* and *E. coli* are sometimes found in drinking water and are known to cause cramping, diarrhea, vomiting, and other medical problems. *Legionella* can live in building water piping and causes respiratory distress. The Centers for Disease Control and Prevention (CDC) monitors waterborne disease outbreaks.

Between 2015 and 2020, the CDC reported 172 waterborne disease outbreaks in PWSs resulting in more than 1,000 people affected. Ninety-three percent of these outbreaks (160) were caused by *Legionella* inside of a building, usually a multistory commercial building where water quality was not maintained. This number of affected water systems is a small portion of the total number of systems in the United States, but waterborne disease outbreaks can occur.

Many of the waterborne disease outbreaks you hear about in the news are related to recreational use—people who get sick from swimming or boating in contaminated waterways. Some illnesses are also contracted from untreated private water supplies.

The waterborne diseases tracked by the CDC are nearly always acute illnesses (having a sudden onset) caused by some kind of biological agent (virus, bacteria, or protozoa), but unwanted chemicals in drinking water can also cause illness, typically over a longer term. That is why chemicals known to have adverse health effects, both acute and chronic (long term, perhaps taking years to see the effect), are limited in drinking water by USEPA standards.

10. Is tap water safer in one area of a community as compared with another?

Rarely. All the tap water from public water supplies must meet all federal, state, or provincial requirements at all points in the distribution system. The condition of the pipes and the flow patterns of water may be different in different areas, and this may cause some differences in water quality, although it usually does not affect water safety. Water at a consumer's tap can also be affected by the condition of the plumbing in that home or business.

11. People are allowed to swim and go boating in our water supply reservoir. Should I worry about this?

Swimmers and boats do add some pollution to water supplies, but it is diluted by all the water in the lake or reservoir and so usually doesn't amount to much. In addition, the water is thoroughly treated before it is delivered to the tap, and the water must meet all USEPA standards. Wildfires, litter, and stormwater runoff can cause far more disturbance to the water quality in the supply reservoir than this kind of pollution.

Still, some water districts do limit the activity on their water supply reservoirs to nonmotorized boats and prohibit swimming and other body-contact activities. Although contamination from gas powered boats and personal watercraft would eventually be removed by the water plant's treatment, some places have put strict restrictions on such craft to prevent any contamination from occurring. Prohibiting activities on a recreational water body can be an unpopular decision, so your water utility would need the support of its public officials if such action were to be taken.

12. Should I use tap water for my baby's formula?

Because tap water meets pathogen standards, most pediatricians consider tap water safe for mixing with baby formula. But you should ask your pediatrician if the water in your particular community is recommended for making infant formula. Although the water utility may meet the standards for some chemicals like fluoride, nitrate, or lead, check with your doctor about whether the levels are appropriate if a baby's sole source of nutrition is formula made with that community's tap water.

For example, water from systems with elevated naturally occurring fluoride should be avoided because of the risk of tooth fluorosis (white lines on teeth). The CDC suggests that it's best to not have all the formula made with fluoridated water to avoid the risk of developing fluorosis.

An American Academy of Pediatrics report recommends regular water testing for nitrates and the breastfeeding of infants in areas where only private well water is available. Water suppliers regularly test for and control nitrates, so most publicly supplied water is safe for making formula for infants.

Testing for lead is also a good idea because lead adversely affects children younger than six years. If you get your tap water from a public water utility, contact them to see if they will test your water for lead. If you get your water from a private well, ask your local health department for the names of certified laboratories in your area that can test your water for lead. Most water utilities will test your home's water for lead, and if it is found, they can suggest a filter to remove the lead.

Check with your pediatrician if you have concerns about using your tap water for formula.

13. Can I use hot water to make soup, coffee, tea, or baby formula?

Hot water should not be used for consumption because hot water tends to pick up metal contaminants from the home's hot water heater and household plumbing.

14. Is tap water suitable for use in a home kidney dialysis machine?

No, not without further treatment. The water used in a kidney dialysis machine comes in close contact with the patient's blood. So, the water needs to be much purer than ordinary drinking water. Aluminum, fluoride, and chloramines may be added to water during the treatment process and are fine for drinking but are not acceptable in water used for kidney dialysis. Because chemicals such as these are critical for the safety of the water, the water must be further treated at the medical center immediately before use in a dialysis machine. Kidney dialysis centers are kept informed about water quality by water suppliers and are best able to advise their patients.

15. Is it safe to take a drink from my garden hose?

It depends. To keep it flexible, a standard vinyl garden hose may contain unhealthy elements, such as lead and vinyl chloride, which can get into the water as it flows through the hose. These metals and chemicals are not good for animals either, so it's not a good idea to fill drinking containers for pets from a garden hose unless the water is allowed to run a while first to flush out the chemicals. If your garden hose or outdoor faucet is NSF/ANSI 61 or NSF/ANSI 372 certified, it means the products meet certain safety standards to be used for drinking water. You can find a list of lead-free plumbing fixtures on www.nsf.org.

Some hoses will not contaminate the water because they are made with a "food-grade" plastic approved by the US Food and Drug Administration (FDA). These hoses will be labeled "safe for drinking" and should be used on recreational vehicles (RVs) when hooking up to a drinking water tap at a campsite. Remember, too, that the opening on any hose could be contaminated by chemicals or germs from lying on the ground or from previous use.

Another reason to avoid drinking from a hose is that older outdoor taps are often not lead-free and thus could add lead to the water.

16. What does NSF/ANSI mean, and why does it matter regarding drinking water?

NSF/ANSI is short for NSF International/American National Standards Institute. These two groups worked together to develop standards and testing for many things, including chemicals and equipment used for drinking water. It's a good idea to look for this group's certification when you're looking for a home treatment system or filter or any kind of plumbing used in drinking water.

17. I collect rainwater to use in my garden. Can I drink that water?

Even though that water is great to use in your garden, it hasn't been treated with a disinfectant at a water treatment plant, so it may have active bacteria or viruses in it. Homeowners often collect rainwater as it runs off their roof for their gardens, but the roof isn't always clean (think of the birds up there!).

18. Is it safe to drink water from a drinking fountain?

Yes, usually. There have been reports in the news that you can be exposed to lead from a certain type of drinking fountain. Some older, floor-standing water fountains were water coolers that had lead-lined storage tanks, and there is a possibility that high amounts of lead could get into the water from these tanks. But, in 1988, these coolers were banned by Congress in the SDWA amendment, "The Lead Contamination Control Act of 1988." This ban was in addition to the 1986 action Congress took to prohibit the use of lead and lead containing solder in any material used in drinking water plumbing. Drinking water from water fountains is generally safe.

19. What's the difference between bottled water and tap water?

Your tap water comes from either a PWS or private well. Public water is regulated by the USEPA to meet quality standards, but private well water isn't regulated. The USEPA requires extensive sampling of water in PWSs. Bottled water is regulated by the FDA, and its standards and testing requirements can differ from those for tap water. Bottled water labels provide details about the water type, and there are many different types of bottled water. Bottled water often contains added minerals or other substances that may not be allowed in tap water.

More questions about bottled water are included throughout this book.

20. Are in-store water vending machines safe?

In most cases, yes. Water from in-store dispensers or vending machines often comes from the municipal water supply via a line that is linked directly to the store's main water pipes. Generally, the water is also treated in some way by the dispenser and is often refrigerated. The most common treatment is reverse osmosis. State health agencies in the United States usually inspect the dispensers on a regular basis, but they are not regulated in Canada. Stores must have policies about maintaining water safety between inspections to make sure the dispenser is carefully sanitized to prevent the possibility of contamination. Make sure the water jugs you are filling at the dispenser are sanitized between uses, as well.

21. Is there salt in my water?

Sodium or salt occurs naturally in drinking water. However, salt also finds its way into water from road deicing, water treatment chemicals, and ion exchange water softeners. Sodium intake from tap water isn't a problem for most US residents, but for those facing heart disease, hypertension, kidney disease, circulatory illness, or a sodium-restricted diet, there are some legitimate concerns. Talk to your doctor if you have concerns about sodium intake.

22. What kinds of things can contaminate drinking water?

The SDWA defines a contaminant in drinking water as a physical, chemical, biological, or radiological substance in water. Some may be harmless, but others pose a health risk.

Drinking water can be affected by microbial contaminants such as bacteria, viruses, or algae, or by chemical contaminants such as lead, nitrate, or chloroform and by newly discovered contaminants such as per- and polyfluoroalkyl (often called PFAS) substances or microplastics. The USEPA's website is a great place to find out more information on the nearly 100 contaminants regulated and those being studied by the USEPA in drinking water.

MICROBIAL CONTAMINANTS

23. What types of living organisms do I need to be concerned about in water?

Microbes occur naturally everywhere in our environment. Not all microbes make you sick, and some are necessary. Pathogens are organisms that cause disease—we usually call them "germs." Common waterborne pathogens of concern are bacteria that can cause illnesses such as botulism, typhoid, dysentery, cholera, and Legionnaires' disease; viruses that cause hepatitis and polio; and *protozoa* that cause giardiasis and cryptosporidiosis.

Utilities treat the water for all these pathogens and more, and they routinely test for harmless coliform bacteria as an indicator organism. Coliforms live naturally in the intestines of humans, and their presence in water could indicate that more dangerous fecal bacteria are also present. Some coliform bacteria, called intestinal pathogenic *Escherichia coli* 0157:H7 by scientists (*E. coli* for short), cause foodborne and waterborne disease outbreaks. Water treatment is designed to remove or inactivate such living organisms in drinking water.

24. How are germs kept out of my drinking water?

PWSs use many processes to treat drinking water and eliminate pathogens.

At most water treatment plants, a chemical disinfectant is added that kills most living organisms in the water. Chlorine gas and its liquid counterpart (sodium hypochlorite) and solid forms (calcium hypochlorite) are the most common disinfectants used in the United States and Canada. The discovery of chlorine's effectiveness at killing cholera and typhoid germs has been heralded as one of the most important health discoveries of the 20th century. Other disinfectants include chloramines (a combination of chlorine and ammonia), ozone, chlorine dioxide, and ultraviolet (UV) light. Your water supplier can tell you what disinfectants are used in your water. This information is included in the water quality report usually found on your utility's website.

Private water sources are usually not disinfected and should be tested annually to uncover possible contamination.

Some organisms, such as *Cryptosporidium*, are chlorine resistant, and water suppliers that use surface water use a multiple-barrier treatment process that includes filtration and, at times, ozone or UV to remove or inactivate these organisms from your drinking water.

25. If I am concerned about my water, how can I kill the germs in it?

Using a timer, bring the water to a full boil on a stove or in a microwave oven and boil it for 1 minute. Because the boiling temperature of water goes down about 2°F (1°C) for each 1,000 feet (305 meters) , people living at high altitudes should increase the boiling time. For example, in Denver, Colo., which is more than 5,000 feet (1,525 meters) above sea level, boiling time should be increased to 3 minutes.

Treating water in this way should be done only in emergencies, because heating and boiling use a lot of energy and create a burn hazard. Always be careful with boiling water! Let it cool in a safe place.

26. Can flu viruses be spread in drinking water?

Treated tap water is not likely to transmit influenza viruses. Conventional disinfection processes that meet current drinking water treatment regulations provide a high degree of protection from viruses. Studies have shown that the H5N1 avian flu (bird flu) virus is killed by chlorine, and there are no known cases of humans getting the flu from exposure to drinking water.

27. What are cryptosporidia? What is cryptosporidiosis?

Cryptosporidium—commonly called "crypto"—is a protozoan parasite that can live in the intestines of humans and animals (hosts). Outside of the hosts, the microbe is protected by a shell called an oocyst, so it is like a seed of a plant, very tough and long-lasting. Once swallowed, the microbe emerges from its shell and infects the lining of the intestine. When this happens, some people get a disease called cryptosporidiosis. The usual time between swallowing this microbe and getting sick is 2–10 days. The major symptom is severe watery diarrhea with cramping abdominal pain that lasts about 10–14 days. Other symptoms can include nausea, vomiting, fever, headache, and loss of appetite. Although cryptosporidiosis is an unpleasant disease, it is not a dangerous one for people with normal immune systems, and not everyone who gets infected gets sick.

Cryptosporidium (plural, cryptosporidia) protozoa gets into the waterways from the stools of infected humans and animals such as cattle, sheep, and wild animals. Rain and melting snow run off over pastures and wildlands and transport the wastes into surface waters, rivers, lakes, and streams. Wastewater treatment does not completely remove these microbes, so infected human wastes also contribute.

In 1993, the water system in Milwaukee, Wis., was contaminated with *Cryptosporidium* oocysts, resulting in more than 400,000 cases of cryptosporidiosis, the largest waterborne disease outbreak in the United States at that time. This event resulted in the USEPA issuing more stringent monitoring, quality, and pathogen removal/inactivation standards for all water utilities.

28. Are all water systems at risk from *Cryptosporidium*?

No. Protected groundwaters that are not mixed with surface water are usually free from these organisms.

29. Is drinking water the only source of *Cryptosporidium*?

No. There are many other sources. Unwashed fruits and vegetables, soil, swimming pools, recreational water, unfiltered stream water, day-care centers, and nursing homes are also common sources. Remember, for all these sources, the common factor is contamination from stools of infected humans or animals.

30. My water supplier found *Cryptosporidium* in the source water. Will I get sick if I drink water from the tap?

If you have a normal immune system, probably not. Because your water supplier was testing the source, they will be extra diligent during the treatment process to ensure that crypto is not making its way to the tap. The multibarrier system utilities have in place will usually remove or inactivate *Cryptosporidium.*

Also, the microbes detected by the utility might be harmless because the test doesn't differentiate between living and dead oocysts. Lots of water is sampled to look for the microbes, much more than would be in a glass of water, so finding a few in that large volume of water doesn't mean every glass of water has crypto in it.

Even then, you would likely be OK. Although all people are presumed to be able to become "infected" from these germs, some people have such strong immune systems that they will not get the disease. For example, estimates have shown that if 1 million people each swallowed one germ, only 5,000 would get infected, and only about two-thirds of these 5,000 would have the symptoms of the illness—a very small risk.

If you or a member of your family is at high risk of infection, however, you should take extra precautions. Infants, the elderly, and immunocompromised people, such as people with cancer, transplant patients, and those infected with HIV (human immunodeficiency virus) or who have AIDS (acquired immunodeficiency syndrome) should follow the advice of their healthcare provider if their water utility reports finding crytosporidia in the water source.

For more information, go to the CDC's website at www.cdc.gov, which has lots of valuable information on this protozoon and the disease.

31. Could my drinking water transmit the AIDS virus?

There is absolutely no evidence that AIDS can be transmitted through drinking water. There is no danger from drinking water for three reasons. First and most important, you can't get AIDS by drinking the virus; it must get into the blood directly. Second, the virus is very weak outside of the body and rapidly becomes noninfectious. Finally, even if present in water sources, the virus is easily killed during the disinfection step of drinking water treatment.

32. Could my drinking water transmit the COVID-19 virus?

No. The coronavirus that causes COVID-19 is easily destroyed by disinfection used by your public water supplier. This virus has not been detected in drinking water.

33. What is *Legionella*?

Legionella is a group of bacteria that can cause illnesses including a pneumonia-type illness called Legionnaires' disease when enough of the bacteria are inhaled. The bacteria occur naturally in waters but multiply in building plumbing and HVAC (heating, ventilation, and air conditioning) systems where the water is warm and there isn't enough disinfectant to kill the microbes. When the tap or shower is turned on, the bacteria are released into the air and can infect building occupants. *Legionella* is regulated by the USEPA through the Surface Water Treatment Rule. This rule requires water utilities to maintain detectable disinfectant residual throughout the distribution system. There is no monitoring requirement for *Legionella*, but the goal is that no *Legionella* be present in the distributed water. This goal is not enforceable. Water in the buildings may not be able to maintain the disinfectant residual if, for example, the water is unused for a long period. The CDC has issued guidelines suggesting that property managers test for the bacteria and make sure that enough disinfectant is in the plumbing to kill the bacteria. Because the bacteria multiply and grow in building plumbing systems, it's important for building managers and homeowners to be aware and manage that risk. The bacteria can also grow in cooling towers, fountains, hot tubs, and humidifiers.

CHEMICAL AND MINERAL CONTAMINANTS

34. Are chemicals found naturally in drinking water nontoxic?

Not necessarily. Many chemicals that occur in nature and seep into water supplies can be harmful to your health. A few examples are arsenic, radium, radon, and selenium. Also, some nontoxic natural chemicals combine with other chemicals to produce harmful chemicals (byproducts). Many naturally occurring chemicals are watched closely by your water supplier, which must test for numerous chemicals regulated by the USEPA.

35. What are organic chemicals? Are they dangerous?

Organic chemicals have mostly carbon atoms connected to hydrogen atoms. A common organic chemical in the home is sugar, so not all organic chemicals are dangerous. Food contains many beneficial organic chemicals essential to our life. You can't tell if something is an organic chemical just by looking at it. For example, table salt looks a lot like sugar but does not contain carbon and hydrogen, so salt is an inorganic chemical.

Dangerous organic chemicals are found in man-made products such as gasoline, cleaning fluid, pesticides, paint thinners, and antifreeze. They are dangerous if they get into your drinking water because many are toxic, such as cancer-causing chemicals called carcinogens.

The USEPA regulates many of these chemicals, and when they are found at levels that exceed standards, water systems install additional treatment to remove these chemicals.

36. Where can I find information about how environmental exposure of chemicals affects human health?

The Integrated Risk Information System (IRIS) database, which is prepared and maintained by the USEPA, contains information about how environmental exposure to various chemicals can affect human health. Although initially developed for USEPA staff who needed consistent information about these chemical substances, the information in IRIS is available to the public on the USEPA website.

37. The movie *Erin Brockovich* focuses on the chemical chromium-6. What is that and is it a threat to other water supplies?

Erin Brockovich is based on a true story about the contamination of drinking water with chromium-6 or hexavalent chromium. Chromium can theoretically occur in six different forms based on its valence or oxidation state—hexavalent refers to its +6 valence state—but usually only hexavalent and trivalent occur in nature. *Erin Brockovich* depicts how the Pacific Gas and Electric Company used chromium-6 as an anticorrosion agent in the plant cooling tower at its facility near Hinkley, Calif. The cooling water was dumped nearby and seeped into the area's private groundwater wells. Hinkley residents reported many health problems, ranging from minor skin irritations to cancer and birth defects. The subsequent legal battle ended in an out-of-court settlement of $333 million for affected Hinkley residents.

The USEPA and the Canadian government regulate the amount of total chromium, which includes hexavalent chromium in tap water, and water utilities regularly test for and control its presence. California regulates hexavalent chromium itself and also regulates total chromium, which is the sum of all forms of chromium.

38. Do hazardous wastes contaminate drinking water?

Yes. As rainwater seeps down through hazardous waste, it can carry hazardous chemicals with it to the groundwater. Some chemicals stick to dirt particles and don't reach the groundwater. Other chemicals, such as cleaning fluid, herbicides, and gasoline, move rapidly down through the ground. Rain can also wash contaminants from a hazardous waste dump into surface waters, which can seep into the groundwater and pollute it. This is one reason there are such strict rules on covers and liners in the areas where hazardous wastes are disposed.

Leaking underground gasoline tanks at gasoline stations and the improper disposal of chemicals (for example, dumping old radiator fluid, metal degreasers, paint thinners, or paintbrush cleaners in the backyard) may also contaminate groundwater.

Hazardous wastes can get into groundwater when they are illegally or inappropriately dumped on the ground by industries. See the previous question that describes what happened when water containing hexavalent chromium was dumped near private wells, and questions later in this chapter that describe what happened when dry cleaning fluid was dumped near private wells. Once groundwater is contaminated with hazardous chemicals, it is very difficult to remove.

Surface water in some regions is also contaminated by chemicals used for deicing roads in the winter. When it rains or the snow and ice melt, these chemicals wash into rivers, lakes, and reservoirs. To prevent this problem, some states post signs on roads that cross watersheds to inform the highway crews not to spread deicing chemicals in these areas.

Improperly treated waste from industrial plants may also pollute surface waters. Many water systems work hard to prevent such contamination, but if it occurs, these systems must treat their water to remove the chemicals.

Pharmaceuticals

39. Are there drugs in my water? How do they get there?

A variety of prescription and over-the-counter drugs (pharmaceuticals) and personal care products have been detected at low levels in source water and some tap water in recent years. This includes small amounts of thousands of products people use every day, from aspirin to sunscreen and bug spray to cosmetics, deodorant, hair products, food supplements, and caffeine. Collectively, medicines and over-the-counter goods are called pharmaceuticals and personal care products, or PPCPs.

Many PPCPs enter the environment through the wastewater streams as they pass through people who use them or are washed off in a bath or shower. Other PPCPs are deliberately flushed down the toilet or poured into a sink. Other drugs and hormones come from agricultural and veterinary use and are found in large quantities in waste ponds at feedlots.

40. Should I be concerned about PPCPs in the water?

Although no one wants to drink someone else's medicine in their glass of water, pharmaceuticals and PCPPs at the low levels found in drinking water are not believed to pose a health risk. Also, improving technology means that increasingly tiny levels of PPCPs can be detected, so although it appears that new contaminants of concern are constantly being discovered, they've been in the water for a long time with no apparent ill effects.

This doesn't mean that there is no need to worry, because more and more people are using PPCPs to feel and look better. Consequently, more of these chemicals end up in wastewater. Your water utility is already monitoring for harmful chemicals and removing them if found, but more health-effects research is needed to determine the long-term consequences of PPCPs. It's important not to flush unwanted or old pharmaceuticals to avoid adding them to the environment. See question 43 for information on safely disposing of medicine.

41. What are endocrine-disrupting compounds?

Endocrine-disrupting compounds (EDCs) are a particular group of chemicals that affect the hormonal (i.e., endocrine) systems of animals. Hormones regulate reproduction and some behaviors, so some PPCPs, such as birth control pills, certain pesticides, and some antidepressants, contain EDCs. Like PPCPs, EDCs can end up in the municipal and industrial wastewater treatment system either through direct discharge into the sewers or via stormwater runoff.

Aquatic life is particularly susceptible to the long-term effects of EDCs because it is constantly exposed to chemicals over multiple generations. More research is needed to determine long-term effects on humans, but thus far, no one has linked drinking water to EDC-related problems in humans.

42. Are PPCPs and endocrine disruptors regulated in drinking water?

No, but some of these compounds have been designated as "contaminants of emerging concern" by the USEPA, which means that the agency is gathering occurrence and health data on these compounds.

43. If flushing drugs down the toilet causes environmental problems, how should I dispose of unwanted medication?

Where available, take the medications to a hazardous waste collection site or take-back program at a medical care facility or pharmacy. Before taking any controlled substance to a collection event, however, check with the organizers to find out if they are authorized to accept the material.

The US Drug Enforcement Administration (US DEA) has an interactive tool on its website (www.dea.gov)

that can direct you to a collection location based on your zip code.

More information about disposing of unused or expired pharmaceuticals is on the FDA's website www.fda.gov.

Arsenic

44. What is arsenic and how does it get in my drinking water?

Arsenic is an odorless and tasteless element that occurs naturally in the Earth's soils, rocks, and minerals. Arsenic also enters drinking water supplies from agricultural and industrial activities. Industrial arsenic in the United States is primarily used as a wood preservative, although its use for preserving wood used in residential projects was stopped many years ago. Arsenic is also used in paints, dyes, metals, drugs, soaps, and semiconductors. Certain fertilizers and animal feeding operations also contribute to arsenic contamination. Copper smelting, mining, and coal burning also contribute to arsenic in our environment.

Groundwater flowing through these deposits can dissolve the arsenic, resulting in elevated amounts of arsenic in well water. Runoff or seepage from minefields, feed lots, and industrial waste sites can also contribute to arsenic contamination.

45. What are the health effects of arsenic exposure?

High doses of arsenic or chronic exposure over the long term can cause thickening and discoloration of the skin, stomach pain, nausea, vomiting, diarrhea, numbness in hands and feet, partial paralysis, diabetes, and blindness. Arsenic poisoning also has been linked to cancer of the bladder, lungs, skin, kidney, nasal passages, liver, and prostate.

46. Is arsenic regulated in drinking water?

The USEPA standard and Health Canada guideline for the maximum allowable level of arsenic drinking water are the same: 10 micrograms per liter (μg/L). This is about one thousand times lower than the amount of arsenic it would take to cause health effects.

If you use a private well, you should consider having it tested for arsenic. Because arsenic has no taste or smell in drinking water, the only way to determine whether it is in your well water is by having a water sample tested by a certified state or commercial laboratory.

The USEPA's website has information on how to find your state's laboratory certification office, and you can then find a list of laboratories certified in your state for drinking water analysis.

47. What home treatment systems best remove arsenic?

Ion exchange, reverse osmosis, and adsorption point-of-use (POU) systems are effective at reducing arsenic levels in most home water supplies. Before installing any treatment system, thoroughly investigate the system's ability to remove arsenic.

Pesticides and Herbicides

48. Are there pesticides and herbicides in drinking water?

Many chemicals, called pesticides and herbicides, are applied to agricultural lands, orchards, and residential lawns to protect them from pests, weeds, and diseases.

For example, atrazine is the most widely used agricultural herbicide in the United States and Canada to control broadleaf and grassy weeds, particularly in corn and soybean crops. It is also used on some residential lawns.

Another example of these chemicals is endrin, which had been used as an insecticide. The use of endrin was banned by the USEPA in 1986, but its presence in drinking water is still regulated.

These chemicals typically enter surface water and groundwater sources by running off crop fields after late-spring or early-summer rainstorms. The chemicals also enter the water from spillage or accidental discharge during production, packaging, storage, and waste disposal. Bottled water can also contain atrazine.

Regarding regulated herbicides and pesticides, studies have shown that these chemicals adversely affect human health at certain levels. For example, the USEPA has set a maximum contaminant level (MCL) for atrazine at 3 µg/L, and Canada has an interim maximum acceptable concentration (MAC) of 5 µg/L for drinking water.

The presence of many of these chemicals, including atrazine, endrin, and others, is regulated in drinking water, which requires water utilities to routinely monitor their source and finished water. If these chemicals are found above acceptable levels, the water utility must remove these chemicals until safe levels are at the tap.

If herbicides or pesticides are found in your public water supplies, information about it will be included in your utility's annual water quality report or CCR, which is delivered to every customer each year and is usually available online.

Surveys reported in the media have pointed out that herbicides and pesticides are present in the water sources in many areas, but in most cases, the levels found are low and water treatment is designed to remove these chemicals at the water treatment plant.

Lead and Copper

49. How does lead get into drinking water?

Lead is rarely present in treated water leaving the water plant or in the water distribution system mains (pipes in the roadways). However, the material in service lines that bring the water from the main in the street to the home or home plumbing can contain lead. In these cases, lead can dissolve into the water that flows through the material and be present at the tap.

The USEPA regulates lead in drinking water at the consumer's tap and requires corrective action by the water utility if high amounts are found. The water supplier must offer to sample the tap water for lead of any customer who requests it.

Canada banned the use of lead pipe and lead-based solder in home plumbing, requiring no more than 0.25% lead in any material in contact with drinking water.

In the United States, lead was banned in pipes and solder in 1986, and in 1996, 2011, and 2022, the ban was extended to include any material in contact with drinking water. Older homes can have lead solder and old fixtures containing lead, however. New materials must be "lead-free" or have no more than an average of 0.25% lead. An NSF or NSF/ANSI mark on a faucet means that the faucet has been tested and meets the standards for "lead-free." Consumers should confirm that any plumbing installations or repairs be done with lead-free material. In 2024, the USEPA set requirements that all lead service lines must be removed by 2037, with a few exceptions.

Public water suppliers in the United States are required to sample several homes they serve for lead and copper at least every three years. The results are posted on their website, or you can call them to find out if lead has been found.

In 2016, high levels of lead were detected in homes served by the PWS in Flint, Mich. The release of lead from service lines (the pipe that runs from the street to the home) was traced to a change to a more corrosive water source and a failure by the water utility to add corrosion inhibitor to the water.

50. How much lead and copper is allowed in tap water by federal requirements?

USEPA regulations require that the 90th percentile of all the results of samples collected in a PWS's distribution system be below 15 µg/L for lead and below 1,300 µg/L for copper. This is an unusual measurement for drinking water contaminants, and it means that 90% of the samples tested must be below that level, which is called an "action level." In 2028, the allowed 90% level for lead will be reduced to 10 µg/L.

51. How does copper get into drinking water?

Public water suppliers must also test customer taps for copper, and sometimes copper is found in tap samples. Some copper occurs naturally in source waters, but usually if copper is found at a home tap, it is because the home service line or home plumbing has copper pipes in it and that copper has dissolved into the drinking water. Copper at levels found in drinking water is usually an aesthetic problem and not a health issue, but an unusually high level of copper (more than 1,300 µg/L) can cause abdominal cramping and diarrhea.

52. Should I remove my lead lines or lead plumbing?

You've been hearing in the news that lead service lines can contribute to lead at your tap and that the Lead and Copper Rule Improvements (LCRI) of 2024 requires that your water supplier remove its portion of the lead service lines by 2037. Some homeowners own the service line from their property line into their home, and some water utilities may only remove that part if the homeowners agree to pay for it. If you think your home might have a lead service line, contact your water utility to find out what your options are. Removing lead service lines is the best way to eliminate the risk of lead release in your water.

If you have your water tested for lead and some is found but you don't have a lead service line, the lead is probably coming from older faucets or from lead solder used to join copper pipes. Talk to your water supplier for advice on certified filters that could remove the lead or a plumber to help find the cause of the lead release. You should periodically clean and replace your faucet aerators, which are found at the end of the faucet. These can trap lead particles that can then be released over time.

53. How can I get lead or copper out of my drinking water?

The first step to take if you think you may have lead or copper in your water is to have the water tested. Contact your water utility to get information on how to get this testing done.

If testing indicates a problem, if your water is corrosive, or if you have rusty water or blue-green stains in your sink, take the following precautions.

If possible, replace your service line and any old lead bearing faucets with a nonlead alternative. (See previous question.)

Whenever water has not been used for a long period—overnight or during the day if no one is home—let cold water run from the faucet for about 5 minutes before using any water for drinking or cooking. Just how long it takes for freshwater from the main to arrive at the faucet depends on your specific location, water pressure, whether you live in a single-family home or an apartment, and so forth. If your home is a short distance from the street, you may be able to run the water for less time. Save the first-draw water for other nonpotable purposes, such as plant watering.

Whole-home treatment equipment that contains lead-removing filters, reverse osmosis systems, and distillation units remove lead dissolved in water that enters your home. However, whole-home treatment devices installed where the water enters the home will not remove lead if it comes from a source of lead in the home plumbing. In this case, a certified filter to remove lead must be used at the tap. Check to see whether the performance of these products has been tested for lead reduction by independent testing and certifying organizations following the methods contained in the appropriate NSF/ANSI standards for equipment used in drinking water applications. More information about these standards for home drinking water units can be found on the NSF International's website, www.nsf.org.

54. What are the health effects of drinking water with lead in it?

Lead may cause a range of health effects, including delays in physical or mental development, particularly in children six years and younger because this is when the brain is developing. For adults, lead can cause kidney problems or high blood pressure, and a recent study has linked lifetime exposure to lead to an increased risk of cardiovascular problems and cataract development in men.

The primary source of lead exposure for most children is ingesting lead-based paint chips and inhaling paint dust in older homes. The USEPA estimates that 10% to 20% of human exposure to lead may come from lead in drinking water. Infants who consume mostly formula mixed with tap water can receive 40% to 60% of their exposure to lead from drinking water. For more information about lead, visit the USEPA's lead information website at www.epa.gov/lead.

In recent years, the USEPA has placed stricter requirements on water utilities for lead monitoring, testing, and treatment; lead service line management; and customer awareness through amendments to the SDWA. The LCRI of 2024 requires the removal of all lead service lines and also aims to reduce lead in tap water in schools and childcare facilities.

Microplastics

55. What are microplastics?

The term *microplastics* represents a broad category that encompasses a wide variety of very small pieces of plastic comprising different shapes, sizes, and chemical types. Although there is no formal scientific definition of microplastics, they are generally understood to be plastic films, fragments, fibers, and pellets smaller than 5 millimeters. In some cases, the term *nanoplastics* is used to more precisely describe the subgroup of microplastics smaller than 1 micron (1000 times smaller than a millimeter). Microplastics can originate from many sources, including the breakdown of larger plastics such as plastic bottles, material shedding from synthetic clothes and other textiles, and industrial discharges, among numerous others, and studies have detected them throughout the environment around the world, as well as in the human body. Ongoing research is being conducted to better understand any adverse health effects, determine the removal efficiency of different water treatment processes, and improve analytical procedures for the detection of nanoplastics.

Nitrates

56. How do nitrates and pesticides get into my drinking water? What health problems do they cause?

Nitrates and pesticides come from fertilizers and pesticides used on farms and home gardens and lawns. Septic tank drain fields and wastes from animal feedlots are other major sources of nitrates. Rainwater takes these chemicals with it and contaminates groundwater and waterways. As rain moves through the soil, microbes in the soil change the ammonia in the fertilizer and drainage from septic tanks and feedlots into nitrates.

The USEPA and Health Canada have set a limit for nitrates because high levels are associated with a rare blood condition in infants, commonly called the blue-baby syndrome (methemoglobinemia). The name comes from the baby's skin, which turns bluish after the baby drinks water containing nitrates because the chemical interferes with the oxygen-carrying capability of hemoglobin in the blood.

If your water supply contains nitrate and comes from a private well, you may want to consider installing a home treatment system that uses reverse osmosis, distillation, or ion exchange. You can contact a local water treatment dealer who can have your water tested and install proper equipment, if necessary, to remove nitrates and pesticides.

Perchlorate

57. What is perchlorate?

Perchlorate is both a naturally occurring and man-made chemical. Perchlorate is used to manufacture fireworks, explosives, flares, and rocket propellant. Perchlorate dissolves easily and moves quickly in underground water and surface water. It breaks down very slowly in the environment. Perchlorate has been found in the water supplies of 4% of PWSs in the United States.

In humans, it can affect iodine uptake by interfering with the normal functioning of the thyroid gland.

58. Is perchlorate regulated?

The USEPA and water suppliers have been monitoring perchlorate in water supplies for more than a decade. The chemical is on the USEPA's Contaminant Candidate List (CCL) but is not currently regulated.

In 2009, the USEPA issued an Interim Drinking Water Health Advisory level of 15 µg/L of perchlorate in drinking water but has not issued a regulation setting a limit in drinking water. The USEPA is required to issue a regulation by the end of 2025.

59. How can perchlorate be removed if it gets in my drinking water?

Some reverse osmosis home treatment units have been certified by NSF International to reduce perchlorate from levels as high as 130 µg/L to 4 µg/L or less in drinking water. However, before installing a home treatment unit, contact the manufacturer to determine if the unit can remove perchlorate from your water supply.

PFAS

60. What are PFAS?

PFAS, or per- and polyfluoroalkyl substances, are a large group of man-made organic chemicals used in many everyday products like nonstick cookware and firefighting foams. These chemicals are concerning because they don't break down easily in the environment, earning them the nickname "forever chemicals." Over time, PFAS build up in water, soil, animals, and even people, making them a persistent problem in the environment and drinking water sources.

A 2023 study by the US Geological Survey estimated that nearly half of US tap water, from both private wells and PWSs, likely contains at least one PFAS chemical. Research suggests that exposure to certain PFAS chemicals may cause health problems, such as issues with reproduction, delayed growth and development in children, a higher risk of cancer, weakened immune systems, and increased cholesterol levels.

The USEPA regulates PFAS to reduce the risks these chemicals pose to human health. In 2024, the USEPA set national drinking water standards for six PFAS, including perfluorooctanoic acid (PFOA) and perfluorooctane sulfonic acid (PFOS), with limits set at 4 nanograms per liter (ng/L) for these two compounds, and 10 nanograms per liter (ng/L) for PFHxS, PFNA and HFPO-DA (generally known as GenX compounds. Further, a Health Index MCL of 1 was set to address the co-occurrence of any of these PFAS compounds. Water utilities will need to test for these six PFAS by 2027 and then take action, such as installing advanced treatment systems, if PFAS levels exceed these limits.

61. How can I remove PFAS from my drinking water?

Some POU devices can remove PFAS. These POU devices use granular activated carbon (GAC) adsorption, reverse osmosis, and ion exchange. If you are concerned about PFAS in your water, only use POU devices that are certified by NSF for this contaminant. NSF doesn't currently certify these units down to the level in the SDWA (4 ng/L), so the unit will only decrease the level of PFAS and possibly not remove it to the USEPA level. If you choose to use a POU device for PFAS reduction at your tap, be sure to understand how often the manufacturer recommends that you replace the filter. More information on these systems can be found on NSF's website, www.nsf.org.

Radon and Radium

62. What is radon, and is it harmful in drinking water?

Radon is a radioactive gas that is dissolved in some ground-waters. It is formed when radium or uranium decays naturally. When inhaled over long periods of time, radon can cause cancer. Experts also think radon has some harmful effects when consumed. However, radon in drinking water generally contributes a very small part (about 1–2%) of total radon exposure from all sources.

The USEPA proposed regulations to reduce the public health risks associated with radon in air and water in late 1999, but final action on the proposal has not been taken. Water suppliers do have treatment methods available for radon removal when and if it is required.

63. My private well has high levels of radon. How can I make sure my drinking water is safe?

The most effective treatment is to remove radon from the water right before it enters your home with a point-of-entry (POE) treatment unit. Two kinds of POE devices remove radon from water: GAC filters and aeration. Before you do this, however, discuss the process and consequences with the proper authorities; after collecting radon, GAC could become a hazardous waste, and aeration puts radon into the air.

64. What is radium, and how is it treated?

Radium is a radioactive element that occurs naturally in some soils and some groundwaters, along with radon. Radium is regulated by the USEPA, and water utilities monitor for it and remove it if it is found in water sources.

Small water systems and private well owners can use ion exchange and reverse osmosis POU devices to remove radium. Water softeners also effectively remove radium, but they do not remove radon. In fact, the radium that is trapped by the water softener may create some radon. Discuss this issue with your water supplier.

65. What are TCE and PCE?

TCE (trichloroethylene) and PCE (perchloroethylene or tetrachloroethylene) are chemicals that have been used in dry cleaning and industrial applications such as metal cleaning and finishing. The use of TCE has decreased, but PCE is still used in some industrial applications. In 2024, the USEPA banned most uses of these two chemicals and currently regulates their levels in drinking water. All water utilities must monitor for their presence.

News stories reported that high levels of PCE and TCE were found in the drinking water at Camp Lejeune in North Carolina during monitoring that was conducted in 1982. The chemicals were traced to leaking storage tanks and a dry-cleaning operation. The contaminated wells were shut down, but concerns remain about the effect of this contamination.

WATER TREATMENT

66. How is my drinking water treated?

There are more than 50,000 public water systems (PWSs) in the United States that serve at least 25 people year-round, or communities. These PWSs are what we usually think of as water systems that serve cities or community water systems (CWSs.) But there are also PWSs serving different kinds of communities such as RV parks, campgrounds, prisons, and schools. These additional 100,000 PWSs are called transient noncommunity water systems (TNCWs) that serve transient populations, such as RV parks where people don't stay very long, and PWSs called nontransient noncommunity water systems (NTNCWS) that serve nontransient populations, such as schools where the same people drink the water at least six months of the year.

Each PWS treats its water based on the unique characteristics of the surface or groundwater it uses as a source. There are many different types of processes that can be used to treat water. How your water utility treats its water is described in the CCR that you can usually find on your water utility's website. Most water utilities use a combination of physical and chemical processes to remove or inactivate contaminants, such as settling, coagulation, filtration, and disinfection. Each of these steps is individually optimized to provide multiple barriers between contaminants in the source water and the customer's tap.

If you get your drinking water from a private well, it usually doesn't have any treatment, and the water is pumped or flows by gravity from the groundwater to your tap.

67. Are all the chemicals in my drinking water bad for me?

No. Some chemicals are good for you, and some minerals are accepted by most to be beneficial in drinking water at proper levels. In addition, many chemicals such as chlorine and phosphate are necessary to improve the quality and safety of the water, and these should have no effect on your health.

Chlorine

68. Is water with chlorine in it safe to drink?

Yes. Chlorine is added during the water treatment process to kill microbial contaminants, and a small amount (called a residual) is left in to maintain water quality in the distribution system. The USEPA limits the amount of chlorine remaining in the water at the tap to safe drinking levels. Many studies have shown that this amount is safe to drink, although some people object to the taste or smell. If this is a problem for you, chill a pitcher of tap water in the refrigerator, or let the water stand in the glass for a short time before drinking, and the taste and smell should dissipate.

69. Is there a link between chlorine and cancer?

Chlorine itself does not cause cancer, but when it reacts with natural substances in the water, like decomposing plant material or chemicals from algae, it can form compounds that might increase cancer risk. The resulting compounds are called disinfection byproducts (DBPs). The USEPA regulates DBPs in drinking water and has set strict limits on the DBP level allowed in treated water. DBPs regulated by the USEPA include a group of four chemicals with the general name trihalomethanes (THMs). Another group includes the sum of five haloacetic acids (HAA5). If your tap water has harmful levels of these, your water supplier will notify you. If you're concerned, contact your water supplier and ask about the levels of THMs or HAA5 in your water.

Remember, disinfection is an important part of the treatment process, and disinfectant levels must remain adequate to kill the germs found in water. Any harmful effects to humans from DBPs are very small and difficult to measure in comparison with the risks associated with germs surviving inadequate disinfection.

70. Should I be concerned about the chlorine in the water I use for bathing or showering?

No. Chlorine does not absorb into the skin and get into your body, and the amount of chlorine in the water is too low to harm the skin itself. The levels in drinking water are also too low to cause problems breathing when the chlorine is released from the water into the air during a shower.

71. What else is used to kill the germs in water?

There are several alternative disinfectants, each of which has its own strengths and limitations. Chloramines are a combination of chlorine and ammonia, and they form very few DBPs. But chloramines are not as strong as chlorine, so they are generally used only as a supplemental or seasonal disinfectant to prevent contamination in the distribution system.

Ozone use is increasing in the United States. It has no taste and is highly effective against *Cryptosporidium*, but water treatment plants must add equipment to handle the highly corrosive and toxic ozone gas. Ozone gas dissipates quickly and doesn't provide protection in the distribution system, so chloramines or chlorine must be added to protect the water all the way to the tap. This is also true for UV light. UV radiation does not form DBPs and is an effective disinfectant in very clear water, but high doses of UV radiation are needed to inactivate large protozoa such as *Cryptosporidium* and *Giardia*.

Chlorine dioxide is a highly efficient disinfectant, particularly in cold water, but it is volatile and hard to store, so it must be made at the water treatment plant.

Aluminum

72. I hear aluminum is used to treat drinking water. Is this a problem? Does it cause Alzheimer's disease?

Aluminum-containing chemicals, called alum (pronounced AL-um) or aluminum sulfate, are used to treat most surface waters. These chemicals trap dirt and then form large particles in the water that settle out; thus, very little aluminum stays in the water. Aluminum is a natural chemical that occurs in many foods, including tea. Even if you live

38

in areas where the level of aluminum in drinking water is much greater than average, your intake from food would be about 20 times your intake from drinking water.

Most research has indicated that the low levels of aluminum found in some drinking water do not contribute to Alzheimer's disease. However, the debate and research continue on whether aluminum contributes to the progression of the illness and whether the contribution of alum in drinking water is a concern.

The USEPA considers an acceptable range of aluminum in drinking water to be between 0.05 and 0.20 milligrams per liter (mg/L). This is a recommendation, not an enforceable standard.

Health Canada also does not have a health-based standard for aluminum but recommends that plants reduce residual aluminum levels in treated water to the lowest possible amount.

Fluoride

73. Is water fluoridation safe?

The results of many scientific studies have found fluoride in drinking water to be safe for consumption. Water utilities that add fuoride must keep the level of fluoride at 0.7 mg/l, and at that level research has shown that no adverse health impacts occur. Fluoride helps build strong, healthy teeth that resist decay. When added or naturally present in the correct amounts, fluoride in drinking water has greatly improved the dental health of US and Canadian consumers since it was first introduced in the mid-1940s. The American Dental Association, American Medical Association, AWWA, US Surgeon General, and numerous other professional groups endorse fluoridation of community water supplies because of the public health benefit it provides. Water fluoridation has been recognized by the CDC as one of 10 Great Public Health Interventions of the 20th century.

By reducing dental cavities, fluoridation has been shown to improve not only the dental health, but also the overall health of children. But at high levels, fluoride does adversely impact health and so the USEPA set an MCL of 4 mg/L in water. At levels higher than 4 mg/L, fluoride can cause fluorosis or staining of the teeth. Recently, there have been limited studies that suggest a risk of slight IQ decrease associated with fluoride at levels higher than those found in drinking water. It's important to note that none of the studies on IQ were conducted in the United States and were instead from areas with high levels of naturally occurring fluoride in water.

More information on fluoride and its benefits can be found on the CDC website, www.cdc.gov.

74. Will I lose the benefits of fluoride in my drinking water if I install a home treatment device or drink bottled water?

Certain types of home treatment devices will remove 85% to more than 95% of all the minerals in water, including fluoride. These are reverse osmosis, distillation units, and deionization units (not water softeners—they leave fluoride in the water). If you use one of these types of devices, consult with your dentist or doctor about fluoride.

The situation with bottled water is less clear. Many bottled waters contain very little fluoride, although a few contain amounts that can be helpful in preventing tooth decay. Bottlers are not required to list the fluoride content in a bottle of water unless it is added at bottling. If the bottled water contains greater than 0.6 mg/L and up to 1.0 mg/L, the label can contain the statement, "Drinking fluoridated water may reduce the risk of tooth decay." Otherwise, you'll have to contact the bottler to determine how much fluoride, if any, is in the water.

TRAVEL

75. If I travel overseas, is the tap water safe to drink?

Tap water generally is safe to drink in many industrialized countries. Many developing countries, however, have inadequate sanitation and water supply infrastructure, so precautions should be taken when visiting these areas.

The CDC recommends that international travelers take precautions before and during travel to avoid potential health problems:

- Contact your physician, local health department, or agencies that advise international travelers at least four to six weeks before departure to schedule an appointment to receive current health information on the countries you plan to visit.
- Obtain vaccinations and prophylactic medications as indicated.
- Address any special needs.

Health Canada also has an active travel advisory site on its website, www.canada.ca/health-canada.

76. Is there something I can take with me when I travel to purify water?

Several portable mechanical water purifiers claim to produce germ-free water one glass at a time. These devices frequently combine a filter and a chemical disinfectant and are available at outdoor equipment stores. If they contain an antimicrobial substance, they must be registered by the USEPA. Some portable purification devices on the market use UV light or other methods that do not involve chemicals and are not required to be registered by the USEPA. Because a traveler cannot determine the proper conditions required to make sure that UV light will kill *Cryptospo-*

ridium and *Giardia*, any device that uses only UV light should be used with caution.

Some hikers and travelers also use purifying tablets or iodine, but these usually add an odd taste to the water and do not protect against *Giardia* or *Cryptosporidium*.

77. When I travel to a different place in this country, sometimes I have an upset stomach for a couple of days. Is this because something is wrong with the water?

If the water comes from a PWS, your stomach problems are likely not a result of contamination problems in the drinking water. However, water with a high mineral content, particularly sulfate, may have a temporary laxative effect if your body is not accustomed to the water. Therefore, the change in mineral content from place to place sometimes does bother travelers for a short time until the body readjusts.

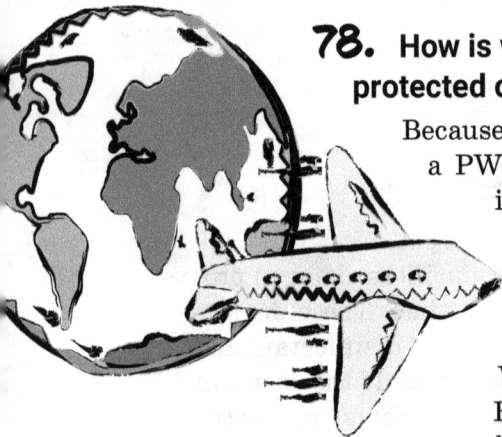

78. How is water quality protected on airplanes?

Because water on US flights comes from a PWS, the water quality on aircraft is governed by the USEPA under the auspices of the SDWA and was specifically addressed in the 2009 SDWA amendment, the "Aircraft Drinking Water Rule."

Following an investigation that revealed contaminated water supplies in the aircraft of several major airlines, the USEPA worked with the airline industry to implement new aircraft water testing and disinfection protocols to protect the traveling public. Health Canada found a similar situation in 2006 and has since been working with airlines to improve onboard water quality. The protocols emphasize preventive measures, adequate monitoring, and sound maintenance practices such as flushing and disinfection of aircraft water systems.

79. We sometimes hear about waterborne illness outbreaks on cruise ships. How is water quality protected on cruise ships?

Unlike outbreaks in a land community, cruise ships provide a controlled environment where it is easy to track the spread of disease; and yes, most infectious disease outbreaks on ships are water- or foodborne caused by unsanitary practices such as lack of hand washing.

That said, most cruise ships operating out of US ports participate in the voluntary Vessel Sanitation Program run by the CDC. CDC personnel conduct unannounced inspections of the cruise ships twice a year and rate the cleanliness of food and water on a 100-point scale. Drinking water items make up 27 of 100 points. Ships can take water from the shore, disinfect it, and put it in tanks on the ship (this is called bunkering). Some ships do, however, make drinking water from seawater. Whatever the source, the water then must be disinfected again before going to the passengers, and records must be kept showing that the disinfection was up to standards.

80. How is drinking water on cruise ships treated?

Cruise ships can bring water from shore, but with the biggest ships using as much as 400,000 gallons (1.5 million liters) of freshwater a day, for some ships this is impractical. Most of the bigger liners have on-board desalination units and use either reverse osmosis or flash evaporators that boil seawater and recondense the steam vapor to produce purified water from the ocean water. This water then goes through a remineralization process to add back minerals

43

that give the water a fresh flavor, followed by disinfection to kill any microbial contaminants.

The wastewater also goes through an extensive treatment process before being discharged.

81. Do I need more water if I'm moving or traveling to a higher altitude?

Dehydration is a valid concern when adjusting to a higher altitude because the altitude can trigger rapid breathing and more frequent urination. Each person's reaction will differ, depending in part on how much altitude change is being made. The basic eight 8-ounce glasses of water should be the minimum intake and more if you are involved in strenuous activities. If you have any special health concerns, it is always a good idea to consult with your doctor before traveling to higher altitudes.

82. Is it OK for campers, hikers, and backpackers to drink water from remote streams?

No. These streams often contain the pathogenic protozoa *Giardia* or *Cryptosporidium* and other organisms that can cause disease.

83. What can campers, hikers, and backpackers do to treat stream water to make it safe to drink?

In addition to or in place of purchasing a water purifier as discussed in an earlier question, any water that looks good enough to drink can be made microbiologically safe by boiling. One minute of vigorous boiling at sea level or 3 minutes at high elevations will kill all germs, disease-producing bacteria, viruses, and protozoan cysts.

Water disinfection tablets made of iodine or chlorine, available at drugstores and camping supply outlets, can be put into a glass of clear water. They take about 30 minutes to dissolve and release the disinfectant. These tablets will not kill *Cryptosporidium* and aren't very effective against *Giardia* (the primary concerns in most areas) or parasitic worms, however, and will not work well in cloudy or colored water. Thus, their use alone is discouraged.

Chapter 2

Taste, Odor, and Appearance

*"Because of you I notice the taste of water, a luxury I might
otherwise have missed."*
— Adrienne Rich, *Like This Together*, 1963

84. Can I tell if my drinking water is OK by just looking at it, tasting it, or smelling it?

Ideal drinking water is crystal clear, has no distinct smell,
no lingering aftertaste, and no mouthfeel sensation such
as a drying feeling or tongue coating. But water can have
an odor, taste, or color for many reasons that do not affect
the safety of the water, including from natural minerals,
such as sulfur or organic compounds. Naturally occurring
tannins, for example, can sometimes be found in water
sources and give water a musty, earthy taste but tannins
don't pose a health risk.

Even if water looks, smells, and tastes good, it doesn't
necessarily mean that it's OK to drink. Untreated water
may contain undetected microorganisms or chemical contaminants that are generally odorless and taste-free but
that can cause disease. Clear, cold stream water, for example, often contains pathogens such as *Giardia*, a parasite
that causes intestinal illness.

Chemicals are a different matter. Most chemicals can't
be seen in water, but many do impart tastes or odors. For-
tunately, the US Environmental Protection Agency (USE-
PA) health limits and proposed limits on regulated chemicals are much higher than the amounts that cause tastes
and odors (T&Os). This means that even if you taste or

smell a chemical odor, the water may still be safe to drink. Still, you should report any problem like this to your water supplier.

The USEPA also sets standards for water's aesthetic qualities that are called secondary maximum contaminant levels (SMCLs). These are not enforceable, but most public utilities abide by SMCLs to control smells, odors, or colors in tap water.

85. What should I do if I notice a change in my drinking water?

Your drinking water should be clear, with little to no taste or smell. If you notice a change in your water, report it to your water supplier right away because it could indicate a problem. Utilities depend on customers to let them know about changes. Although water may leave the treatment plant without any taste or odor issues, these can sometimes develop in the pipes on the way to your home.

86. Why does my drinking water taste or smell "funny"? Will this smelly water make me sick?

Our senses of taste and smell help protect us against unsafe foods and beverages, but T&O problems with your water are usually harmless. Some contaminants that could affect your health can be tasted in drinking water (e.g., gasoline, oil, and ethylene glycol), but some harmful contaminants (such as lead) cannot be tasted. T&O issues occur for a variety of reasons, including those detailed in the following sections.

Chlorine smell

A noticeable flavor (or odor) in water can come from the chlorine that is added by your water utility to kill bacteria and viruses. If you get your drinking water from a public water system (PWS), the utility is required by the USEPA to make sure that every tap in their systems has some level of chlorine (or chloramine) in the water to protect the water once it is in the distribution system. However, if it seems unusually strong, contact the water utility to see if there's a problem.

If you get your drinking water from a private well that isn't treated and you smell chlorine, that would indicate a problem. You should contact your local health department.

Rotten-egg odor

A rotten-egg odor in some groundwater is caused by a nontoxic (in small amounts) and smelly chemical—hydrogen sulfide—dissolved in the water. This is a common issue with many private wells, and anyone who has visited natural hot springs knows the smell. A rotten-egg odor may be a sewer smell; if you are uncertain, have a plumber check it out.

Earthy, musty odors

As some algae, bacteria, and fungi grow in water, they give off nontoxic, odorous chemicals that can cause unpleasant odors in drinking water. Different microorganisms can produce different T&Os; the two most common ones in drinking water are identified as "earthy" and "musty." Although these T&Os are unpleasant, the chemicals producing them, including geosmin and methylisoborneol (MIB), are not harmful to humans.

Metallic taste

Metallic tastes can come from copper and iron dissolved from the pipes carrying drinking water from your public water supplier. Iron can also be found in well waters at levels that can give the water a metallic taste. Iron generally has no effect on health at the levels that could be consumed through water. Higher levels of copper can cause short-term health problems like diarrhea and cramping. Although zinc and lead don't cause a metallic taste in water, they can also dissolve from pipe materials in the same way and possibly at the same time as iron and copper. Zinc and lead can affect your health. Zinc at high levels can cause abdominal problems and even at low levels, lead can cause serious neurological problems. If your water has a metallic taste, it would be a good idea to ask your water utility how to have it tested for all metals, and if you are on a private well, you should contact the local health department for their advice on testing.

Sewer smell

Sewer-like smells in your water can be caused by a cross-connection with a wastewater collection system component and could be a serious problem. Have a certified plumber inspect your plumbing for proper connections and valves to prevent this from happening.

Flat-tasting water

Water tastes flat if it lacks oxygen or minerals. Water that has been treated with reverse osmosis and distilled water can taste flat to many people, as can stored or boiled water. In fact, some bottled water suppliers add minerals to their water to prevent it from tasting flat.

87. What can I do if my drinking water tastes "funny"?

As mentioned in the previous question, if you notice a change in the way your tap water looks, smells, or tastes, the first thing you should do is call your water supplier.

Ways to remedy the problem of "funny-" or undesir-able-tasting water include the following:

- Store drinking water in a closed glass container in the refrigerator because cold drinking water has less flavor than warm drinking water. Although some hard plastic bottles such as those with the symbols HDPE (high-density polyethylene), PET (polyethylene terephthalate), or PP (polypropylene) on the bottom are OK for storing drinking water in the refrigerator, other types of plastic will cause a taste in water and shouldn't be used.

- Flat-tasting water will taste better if you pour it back and forth between two clean containers before drinking it. This will put oxygen from the air back into the water and remove the flat taste.
- Add 1 or 2 teaspoons of lemon juice to refrigerated drinking water.

If your tap water from a private well or your water consistently has T&O problems, consider installing a point-of-use (POU) water treatment product that has been tested by an independent organization following ANSI/NSF 42, and 53, the standards that adddress equipment used in drining water. More information on this certification can be found on NSF International's website, www.nsf.org.

POU devices often contain activated carbon (sometimes mistakenly called activated charcoal or just charcoal), which can remove many T&O-causing chemicals, including chlorine. Some devices use reverse osmosis (forcing water through a membrane that holds contaminants back) or distillation instead of activated carbon. POU devices can be tap mounted or located under the sink. If you wish to treat all the water entering your home, you'll need to install a point-of-entry (POE) device at the main service line coming into your house. If you plan on storing water from a home filtering device, store it like you would a food, using clean, airtight containers, and refrigerate it.

If the problem is a rotten-egg odor, consider installing home treatment equipment that will remove hydrogen

sulfide. First, however, make sure the odor is not coming from the hot water heater or the drains in the sink. You can best do this by filling a glass with cold water and sniffing for any odors in an area away from the sink.

All home treatment devices require maintenance and can be expensive to install and operate. This added treatment is the responsibility of the home or business owner, so be sure you have a plan to maintain the system.

Contact your water supplier with your concerns if you receive your water from a PWS. Water supply system employees are usually aware of current T&O problems in their systems and are ready to answer customer questions. Your call may alert them to an emerging problem, so report any unusual taste or odor to your water supplier.

88. Why do the ice cubes from my freezer used to cool my water make the water taste funny?

This is a common complaint that has no single, simple explanation, but the off flavors are generally absorbed from food or other materials in the freezer. Many items in a refrigerator and freezer can give off odors that are absorbed by the ice. If you have an automatic icemaker, harmless bacteria can grow in the water feed line and cause odors. Smelly substances in use near a freezer may be absorbed into the ice. Though annoying, these "off flavors" are generally not harmful and can sometimes be lessened by cleaning and defrosting your freezer and ice cube trays more often. To help prevent odors in the freezer altogether, put an open box of an odor absorbent, such as sodium bicarbonate (baking soda), in the freezer and refrigerator units.

89. White stuff appears in the glass as ice cubes from my freezer melt. What is the white stuff, and where does it come from?

Ice cubes freeze first on the outside, so the center of the cube is the last to freeze. As an ice cube freezes, the dissolved minerals in the water are pushed to the center and are trapped there. Near the end of the freezing, when not much water remains in the center of the cube, the minerals become very concentrated and form the white stuff, which is technically called precipitate. This precipitate causes the white stuff in your glass and is not toxic.

90. Why does drinking water often look cloudy when first drawn from a faucet?

Cloudy water can be caused by tiny air bubbles in the water similar to the gas bubbles in carbonated soft drinks. After a while, the bubbles rise to the top and are gone. This type of cloudiness occurs more often in the winter, when the tap water is cold and holds more air. When the tap water is drawn into the house, it is warmed in the plumbing.

Another cause of cloudiness in cold water comes from calcium. In certain waters, the nontoxic insoluble (hard to dissolve) chemical calcium carbonate will precipitate when it is cold. Because it is white, this precipitate can cause the water to look cloudy, but the particles will settle to the bottom of the glass (usually in about 30 minutes) in contrast to the air bubbles discussed previously that quickly rise to the top of the water. Water with calcium carbonate precipitate in it is perfectly safe to drink or use for cooking, though it may look unappealing.

91. My drinking water is reddish or brown. What causes this?

Most reddish-brown color found in tap water is nontoxic, but it can be a nuisance. It can stain fixtures and clothing in the washer, and it looks bad. There are several possible causes.

Natural organic matter produces tannins in many well waters and can give the water a dark brown color. Known as humic acids, they are caused by decaying organic substances in soil and leaves and can also be found in peat, coal, many upland streams, and ocean water.

Tannins also come from decaying vegetation that leaches its color into the source water, like the way water changes color after tea leaves are added to it. Typically, this color must be removed by the treatment plant, although some POE devices on the market claim to filter out color.

Iron, which is common in nature, may be dissolved in your drinking water. When iron is dissolved in groundwater, it is colorless, but when it combines with air as water is drawn from your faucet or elsewhere in the system, the iron turns reddish-brown. This can happen before your eyes after a glass of tap water is drawn.

The drinking water pipes in the street leading to your home or in your home may be rusting, creating rusty-brown water. Galvanized pipes could be one source. Also, your hot water tank may be rusting. Corrosive water can cause this type of problem. If you are having trouble and your neighbors are not, then your own pipes or water heater are probably the cause of the problem. When your plumbing is rusting, lead and copper may be getting into your drinking water, as well. This is important, so call your local water supplier or have a plumber check your plumbing. You might have an electrical circuit grounded improperly to your water pipes, which induces corrosion in the pipes. Letting the water run a while will often clear the water (save the rusty water for plants). To avoid problems with lead and copper, the USEPA has established regulations that limit the amount of lead and copper allowed in your water, and all the water suppliers by law must make sure that drinking water is not corrosive and therefore doesn't dissolve metals from metal pipe.

92. My drinking water is dark, nearly black. What causes this?

When manganese, a naturally occurring chemical, dissolves in groundwater, it is colorless. When manganese combines with chlorine, and in some cases, is exposed to air, it turns brown or black. The USEPA established a recommended limit for manganese in drinking water because of this dark water aesthetic problem. Many water treatment plants have processes to eliminate the manganese before it reaches your tap.

You may have heard that manganese is getting more attention as a contaminant of concern in drinking water. That's because at higher concentrations, it can have neurotoxic effects. The USEPA issued a Health Advisory for manganese in drinking water, recommending that people limit their intake of manganese to less than 0.3 milligrams per liter (mg/L) for infants and less than 1.0 mg/L for the general population, but the USEPA hasn't set an enforceable limit based on health effects. Health Canada set a limit in drinking water of 0.12 mg/L.

If your drinking water is dark or turns black, you should report your problem to your water supplier. A POU filter might solve the problem, but be sure it is NSF-certified for manganese removal. Look at NSF International's website for more information before purchasing a POU device at www.nsf.org.

Chapter 3

In the Home

Water, like religion and ideology, has the power to move millions of people. Since the very birth of human civilization, people have moved to settle close to it. People move when there is too little of it. People move when there is too much of it. People journey down it. People write, sing and dance about it. People fight over it. And all people, everywhere and every day, need it.
— Mikhail Gorbachev, quoted in Peter Swanson's
Water: The Drop of Life, 2001

93. How much water does one person use each day?

Most people in the United States drink about 8 cups, or a half of a gallon (1.9 liters), of water per day. The US average is nearly 82 gallons (310 liters) used each day per person for all home uses. Total per-home water use includes outdoor uses, such as lawn irrigation, along with indoor uses like washing clothes and dishes, as well as flushing toilets and taking showers or baths.

Because of other community uses (e.g., industrial, commercial, schools, parks, and firefighting), your water supplier treats and pumps much more water than is used in households. A recent study of 1,100 US water suppliers showed that to supply all the water needed for all uses, the average amount of water pumped is more than 100 gallons (378 liters) each day for each person.

In Canada, the average total home water use is similar to US home use.

94. How long can I store drinking water?

Drinking water that is thoroughly disinfected, such as water from your public water supplier, can be stored for at least six months in capped containers, such as plastic or glass, that will not rust or break. Make sure the storage container is completely cleaned before filling. Water that has been boiled for 1 minute, or 3 minutes at high altitudes, can be stored for up to one year.

Bottled water should be stored unopened in a cool place; under warm conditions, the water may taste like the plastic it is stored in because plastics sometimes leach chemicals. Replace the water every six months and keep it sealed; this will also minimize the "flat" taste that occurs after extended storage. Keep stored water out of the direct sunlight and away from other stored chemicals.

If possible, store water in a refrigerator to help control bacterial growth. Water is not sterile or devoid of living things, but it should be safe from harmful microorganisms. The chlorine residual from your tap water might only last about a month or less in stored water.

95. What is food-grade plastic, and how can I tell if my bottle is safe to use to store drinking water?

The Food and Drug Administration (FDA) has designated some plastic types as being safe for contact with food. These include HDPE (high-density polyethylene), LDPE (low-density polyethylene), PP (polypropylene), and PET (polyethylene terephthalate). Look for these symbols on the bottom of the container. Sometimes there is a "food-safe" symbol that looks like a glass and fork. If the plastic isn't food grade, it could leach chemicals into the water.

96. How much water should I store for emergencies?

A good rule of thumb is to store 1 gallon of water per person per day for consumption. Emergency planning experts recommend storing enough water for at least three days, which means a family of four should store about 12 gallons (45 liters) of water. People with special needs, such as nursing mothers, young children, and family members and pets with illnesses, may require more water to be available.

97. How do I treat my water in an emergency?

If the water has been contaminated by living organisms, the best way to disinfect your water is by boiling it. Bring the water to a boil for 1 full minute (3 minutes if you are at a high altitude), then allow it to cool before storing. Be careful not to burn yourself!

When boiling is not practical, you can chemically disinfect your water. Commonly used household chemical disinfectants are chlorine and iodine. Chlorine can be found in common household bleach. Make sure the bleach does not have additives. Check the label: If the available chlorine in the product is around 1%, add 10 drops to 1 quart or liter of water; if it's 4–6%, add 2 drops to 1 quart or liter of water; if it's 7–10%, add 1 drop to 1 quart of water. If the percentage is not listed, add 10 drops to 1 quart or liter of water; double the number of drops if the water is colored or cloudy. Mix the water thoroughly and allow it to stand for 30 minutes. After this time, the water should have a slight chlorine odor; if not, repeat the chlorine addition and allow the water to stand an additional 15 minutes. If the chlorine taste and

odor (T&O) is too strong, let it stand exposed to the air for several hours or pour it back and forth from one clean container to another to aerate the water and reduce the remaining chlorine. If sediment settles to the bottom, pour off the top layer of water and leave the sediment behind.

Common household iodine from a medicine cabinet or first-aid kit may also be used. Add 5 drops of 2% tincture of iodine to 1 quart or liter of water. If the water is colored or cloudy, add 10 drops. Let the water stand at least 30 minutes before use. Iodine tablets are also available from outdoor outfitting stores because they are sometimes used by backpackers and campers to disinfect untreated water. These stores also have portable filters.

Note that chlorine and iodine are only somewhat effective in protecting against *Giardia* cysts and may not be effective at all against *Cryptosporidium* oocysts. Therefore, use chlorine or iodine only on groundwater supplies (wells), which are not likely to have these contaminants. Water from rivers, lakes, reservoirs, or springs needs to be boiled. A good item to have on hand for emergencies is a ceramic filter, such as those used for camping and backpacking. These filters are effective in removing *Giardia* cysts and *Cryptosporidium* oocysts.

If your water service stops during an emergency, remember that the water in your hot water tank, melted ice cubes, and the water in your toilet tank reservoir can be used as long as this water was collected before the emergency occurred. This water can be used for nonpotable uses such as toilet flushing and general washing, but it shouldn't be consumed. Be sure to discard any ice cubes that might have been made during the time your tap water was unsafe!

Boiling and other disinfection practices, however, do not remove chemical contaminants.

98. Should I use hot water from the tap for cooking?

As noted in question 13, cold water is best to use. Hot water heaters and plumbing are not designed to preserve water quality. Water in hot water heaters loses its chlorine and promotes the growth of bacteria. Hot water is also more likely to contain rust, copper, and lead from your household plumbing and water heater because these contaminants generally dissolve faster into hot water than into cold water.

Regarding hot water, insulating your hot water pipes will help the water in them stay warm between uses. So, after the first use of the day, hot water will come to the tap sooner, thus conserving water. Hot water should be kept between about 110 and 120°F (43–49°C) to prevent the growth of microorganisms that could be a health issue while reducing the chances of scalding or burns.

99. Is it OK to use hot water from the tap to make baby formula?

No. As noted in questions 13 and 98, hot water may contain impurities from the hot water heater and plumbing in your home. Only cold tap water should be used for making baby formula. It's always a good idea to check with your pediatrician about the best way to make baby formula.

100. Is it OK to heat water for coffee or tea in a microwave in a foam cup?

Not unless the cup is labeled as "microwave-safe." Look for a small symbol on the foam that looks like a microwave with wavy lines. That's the FDA's symbol for "microwave-safe." Common foam cups can leach harmful chemicals when heated.

101. Is water that comes out of a dehumidifier safe to drink?

No. Dehumidifiers suck water from the air that has not been disinfected. Bacteria and mold spores can accumulate in this water. Also, the condenser coils in many dehumidifiers are copper, which could leach into the water. Dehumidifier water is fine for other nonpotable uses, such as watering plants, adding to car batteries, or using in steam irons.

102. How is the water from my refrigerator door different from my tap water?

Your refrigerator should be hooked up to your home's main water supply line, so it is receiving the same water as comes from the tap. However, refrigerator water usually goes through a point-of-use (POU) filter and is chilled in the refrigerator's cooling compartment before being dispensed. As with any POU filter, check with the manufacturer to determine what substances are removed from the water and how often the filter needs to be changed.

HOME PLUMBING

103. What should I do to avoid cold-weather problems with my pipes?

Outdoors, close all valve connections to the outdoor spigots. Disconnect and drain all outdoor hoses. After detaching the hose, remove any nozzle attachments, and open the faucet to drain the water from the pipes. Insulated caps, available at home improvement stores, can be affixed to the exposed faucets. Use an air compressor to blow remaining water out of underground sprinkling systems only after the valve to your potable water has been closed to avoid air from going back into your house. Close and insulate foundation vents that are near water pipes.

Indoors, insulate pipes or faucets in unheated areas, such as basements or garages, and under your kitchen or bath cabinets as needed. Leave cabinet doors open on extremely cold days to allow the warmed household air to heat the pipes. Consider locating a stand-alone (space) heater near areas known to be susceptible to freezing if there is an extended cold snap. Make sure you have access to your master valve in case pipes do freeze and rupture, and educate everyone in your household on how to shut off the water. If you are expecting severe cold weather and are worried about your pipes freezing, you can also leave a steady, fast drip of water flowing from the tap during the worst of the cold spell. But remember that any extra flow is a waste of water.

Also, check with your local water company; you may be responsible for keeping the meter from freezing, as well. In other places, meters are maintained by utility personnel.

104. Where do I find my home's master valve?

The most common locations for the master valve in your house are

- where the water supply enters your home,
- near your clothes-washer hookup, and
- near your water heater.

To determine if the valve is the correct one, turn it off and see if it shuts off all water faucets in your home. If not, repeat this process with each valve you find until you identify the correct one. If you are unable to locate it or it is inoperable, contact a plumber for assistance. Once you've found the valve, mark it with something distinctive, like bright paint, a tag, or ribbon, so you can locate it quickly in case of an emergency.

If you live in an apartment, have your apartment superintendent show you how to locate and shut it off if needed.

105. How can I tell if I have leaks in my home plumbing system?

To check for leaks in your home, first determine whether you're wasting water, then identify the source of the leak. Look at your bill to determine your water usage during a colder month, such as January or February when you won't be using outdoor water for irrigation. Be sure to look at the amount of water used, not just the cost of the water. If your household has four members and your usage exceeds 12,000 gallons (45,400 liters) for one month, you probably have some serious leaks. Other ways to determine leakage, according to the US Environmental Protection Agency (USEPA), include the following:

- Check your water meter before and after a 2-hour period when no water is being used. If the meter changes at all, you probably have a leak.
- Identify toilet leaks by placing a drop of food coloring in the toilet tank. If any color shows up in the bowl before you flush, you have a leak. (Be sure to flush immediately after the experiment to avoid staining the tank.)
- Examine faucet gaskets and pipe fittings for any water on the outside of the pipe to check for surface leaks.

You can fix many leaks by replacing worn washers and gaskets, tightening and taping showerheads, and replacing toilet flappers.

Some utilities will conduct a free water audit of your home or provide you with a printed survey to help you detect leaks and find ways to conserve water. Contact your utility for more information.

106. How can I tell if my toilet is leaking?

To check, put a few drops of food coloring in the tank, wait about 15 minutes, and look in the bowl. If the food coloring shows up there, the tank is leaking and should be fixed. Toilets should be checked for leaks every year. Some utilities give away conservation kits with food coloring tablets, flow restrictors, and plastic bags to fill with water and put in toilet tanks.

107. Why are there aerators on home water faucets?

As the name suggests, aerators add air to water as it flows out of the tap. When mixed with water, the tiny air bubbles prevent the water from splashing too much. The added air also improves the taste of the drinking water, making it less flat. And because water flow is diminished through an aerator—often reducing it to half the regular flow—aerators also help conserve water. Aerators should be removed and cleaned regularly because they can accumulate particles and microorganisms.

108. Why do hot water heaters fail?

All waters have natural corrosive properties, so water heaters will eventually rust from the inside to the outside. The time it takes for this to happen varies depending on how corrosive your water is, what has been done to mitigate the corrosiveness, and the original quality of the water heater. Because corrosion is also a factor in water pipes, many water suppliers take steps to decrease the corrosiveness of the water they deliver.

Check with the owner's manual for your specific water heater for the manufacturer's maintenance recommendations.

In hard-water areas, the harmless minerals causing hard water tend to form a scale at the bottom of the hot water heater that can result in the failure of the unit. Occasionally, flushing the water heater from the bottom will prevent some, but not all, of the scale from forming. Using a water softener should minimize this buildup. Also, iron and manganese can accumulate in the bottom of the heater. Thus, you should drain the water heater periodically to remove scale, rust, and mineral particles. However, turn off the heating element for a while before draining to allow the water to cool; if this is not possible, use care because the water is hot, or have a plumber do this for you.

109. What causes the banging or popping noise that some water heaters, radiators, and pipes make?

Each noise has a different cause. In a water heater, some nontoxic minerals in the water form a rough coating on the inside of the heater when the water warms up. When the container walls are rough, air bubbles form as the water heats. When these bubbles burst, they cause a popping noise. A new water heater will have smooth walls without scale buildup, so the water will not form bubbles as it heats up, and no noises will be heard. In a home radiator heated with steam, the banging noise is caused by the condensed water (steam) pooling in the bottom of radiator. Open the inlet valve all the way so the water can run out to eliminate this.

Pipes make noises for two reasons. First, when you open a hot water tap after water hasn't been used for a while, the pipe leading to the tap will be cold. As the hot water runs through the pipe, the pipe heats up and expands. This will sometimes cause the pipe to creak or make similar noises.

The other reason is water hammer. When water—hot or cold—is moving fast through a pipe and the flow is stopped quickly, the water keeps moving for a while, like a train plowing forward during a wreck. The moving water finally bangs against (hammers) a faucet or valve, making a loud noise, like a hammer hitting

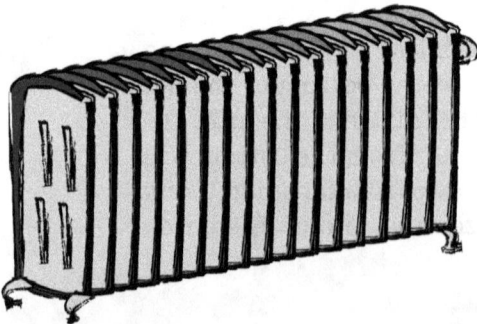

metal. If you've noticed this problem in your home, it can easily be corrected by turning the water off more slowly, installing small standpipes on the affected pipes, or padding the pipes at appropriate locations where it contacts another object to reduce loose pipe rattling.

110. There is a blue-green stain where my water drips into my sink. What causes this?

Blue-green stains come from copper, which probably leached from your home plumbing and dissolved in the tap water. The conditions that cause copper in the water also can introduce lead into drinking water, and high amounts of either lead or copper can cause health problems. Call your local water supplier to discuss what actions can be taken to reduce the level of copper and lead in your water. If your water is from a private well, have your tap water tested for lead and copper. To clean the sink, check with your local hardware store for stain-removal products.

HOME TREATMENT

111. Should I install home water treatment devices?

Water delivered by a public water system must meet stringent requirements for chemical, physical, and biological contaminants that could cause illness, and home water treatment is generally not needed to meet federal, state, or provincial drinking water standards.

In some small communities, installing sophisticated treatment technologies to meet the regulations may be too expensive for the ratepayers to bear. In these cases, in-home water treatment systems may provide a low-cost alternative. The USEPA has authorized the use of home treatment equipment to meet the requirements of the Safe Drinking Water Act (SDWA) regulations for some contaminants. If this is the case in your community, your water utility may provide you with a POU or point-of entry (POE) unit or information on how to obtain, install, and maintain a POU or POE device.

Otherwise, use of POU/POE equipment is a personal decision. If you draw your water from a private well, you may want to consider installing a POU or POE home water treatment device. If you are concerned about aesthetic qualities such as color, taste, odor, or hardness or desire additional barriers against disinfection byproducts (DBPs) or residual traces of contaminants such as lead or *Cryptosporidium*, you might consider a home treatment unit. Be sure that you read the information about the product carefully, or talk to a qualified distributor, so you get the right product for the concern that you have and check to ensure that the device is certified by NSF International (see www.nsf.org).

If you decide to use a home water treatment device, you must be careful to maintain it properly according to the manufacturer's instructions. For example, modern POU and POE treatment products sometimes shut off automatically when a filter is depleted or warn users when replaceable cartridges need changing or routine maintenance is needed.

112. How do I know which type of POU/POE to use?

POE systems are installed at the service line where the water comes into the house, and so they treat all the water used in the home. POU systems treat smaller amounts of water and can be put in several places within the home: countertop, faucet-mounted, under-sink cold tap or under-sink line bypass, and even on showerheads. Treatment units can be grouped into six general categories:

Particulate filters can remove particles, including lead, iron, and manganese particles, of different sizes.

Adsorption units usually contain activated carbon (sometimes incorrectly called activated charcoal or just charcoal) and may remove chlorine, T&Os, and organic compounds such as DBPs. Microbes can grow in these units, but by following the manufacturers' guidelines for periodic maintenance, microbes can be controlled. Adsorption units are generally not designed to remove copper and lead. Certain special filters will remove dissolved lead, but check their claims with independent organizations, including NSF International.

Oxidation/filtration systems change iron (clear water turning red) or hydrogen sulfide (the rotten-egg odor) into a particulate form that can be filtered out of the water. These are primarily POE systems and are frequently used by people who have their own source of water, such as a private well.

Water-softening systems soften water by trading (exchanging) nontoxic calcium and magnesium that cause hard water for other nontoxic chemicals that do not cause hardness, such as sodium. These units must be renewed (regenerated) periodically with salt. Sodium levels in the softened water will be higher than in the unsoftened water, which may be a concern for some health conditions. It is possible to plumb the softening system to leave a source of unsoftened water (the cold water tap in the kitchen, for instance) for drinking and cooking.

Reverse osmosis units remove hardness; chemicals such as nitrates, sodium, dissolved metals (such as lead and copper), and other minerals; and some organic chemicals. Reverse osmosis units also remove fluoride. Some units are sensitive to chlorine, so a chlorine-removal step is usually included before the reverse osmosis unit. Reverse osmosis units do allow some organic chemicals to pass into the treated water, however. Therefore, sometimes these systems are followed by adsorption units to remove these organic compounds. Reverse osmosis units are usually POU devices that treat relatively small volumes of water. These units require high pressure to function properly, so be sure to check that the unit will work on your home plumbing. Remember that if the equipment removes the disinfectant in your tap water, the water must be stored properly to avoid subsequent microbial contamination.

Distillation units boil the water and condense the steam to create distilled water, removing some organic and inorganic chemicals (such as hardness, nitrates, chlorine, sodium, and dissolved metals). Distillation units also remove fluoride. However, some organic chemicals may pass through the units with the steam and contaminate the distilled water unless the unit is specifically designed to avoid this problem.

All POU and POE units require maintenance and should be bought from a reputable dealer. Their performance should be tested and validated against accepted standards like those used by NSF International, the Water Quality Association, and UL Solutions. These standards allow manufacturers' performance claims for drinking water treatment unit products to be tested and checked. You can investigate claims made for any unit by visiting the website of the testing organization named in the product's claims :

- NSF International: www.nsf.org
- Water Quality Association: www.wqa.org
- UL Solutions: www.ul.com

The USEPA has an Environmental Technology Verification (ETV) program that has tested numerous products for their effectiveness in removing certain contaminants. See www.epa.gov.

113. I heard about a water treatment device that uses an electromagnet to treat water. Does this work?

No credible scientific study has supported this claim.

114. Is distilled water the "perfect" drinking water?

Because distillation involves boiling the water, distilling removes harmful bacteria and viruses, *Giardia* cysts and *Cryptosporidium* oocysts, and many nuisance minerals, as well as harmful chemicals like lead, copper, nitrates, sodium, some organic contaminants, and chlorine. But distillation also removes the water's natural minerals, leaving the water with a "flat" taste. Additionally, some organic contaminants, like chloroform and cleaning fluid (solvents), may leave the water with the steam and end up in the final water when the steam is cooled, so most water distillers have added treatment to prevent any organics in the steam from ending up in the final product.

Many people keep a store-bought container of distilled water for use in steam irons and car batteries and for watering plants (water from dehumidifiers can also be used for this purpose), and because most of the minerals are missing, using distilled water to make tea or coffee will avoid a buildup of white calcium–magnesium scale on the kettle or pot. Using a water softener will also take care of this problem.

However, except in special cases, such as removing salt from seawater to make drinking water, distilled water is too expensive for your public water supplier to treat large volumes of water, and it is really not necessary to treat your water to such an extent. There can also be health risks with long-term consumption of water lacking in natural minerals, so speak to your doctor before going this route.

HARD WATER AND SOFTENING

115. What is "hard" water?

Hardness in drinking water is caused by calcium and magnesium—two nontoxic, naturally occurring minerals in water. If calcium or magnesium is present in your water in substantial amounts, the water is said to be hard because lathering soap for washing is hard to do, and cleaning with hard water is inefficient. Water containing little calcium or magnesium is called soft water and is better for laundering and other purposes.

Other indications of water that is excessively hard include towels and clothes that look dingy and feel harsh and scratchy after laundering; spots on dishes and glasses after they've been washed; filmy shower doors; and sticky, dull hair, even after washing. A buildup of mineral deposits can also reduce water flow in domestic pipes, and faucet aerators may become plugged if not cleaned often.

116. Should I install a water softener in my home?

If very hard water is interfering with your laundry or personal and home grooming, a water softener can help. You can find out the degree of hardness of your drinking water by contacting your water supplier or by testing it yourself with a home water test kit. The hardness of your water will be reported in grains per gallon, milligrams per liter (mg/L), or parts per million (ppm); one grain of hardness equals 17.1 mg/L or ppm of hardness. The higher the hardness number, the more a water softener will help.

USEPA Hardness Scale

Classification	mg/L or ppm	Grains/gal
Soft	0–17.1	0–1
Slightly hard	17.1–60	1–3.5
Moderately hard	60–120	3.5–7.0
Hard	120–180	7.0–10.5
Very hard	180 or greater	10.5 or greater

A water softener can reduce the formation of scale in your hot water system and make washing easier. Water softeners replace the calcium and magnesium with sodium or potassium, which dissolve in water and do not leave deposits.

Water softeners are regenerated with salt, either table salt or potassium chloride (KCl). After the salt is used, it goes down the drain and into the environment via a wastewater treatment plant. It's best to regenerate a softener after a set amount of water has passed through it rather than on a particular time schedule to avoid wasting salt by regenerating too soon or using the softener after it has stopped softening. Manufacturers' directions should provide this information.

You may consider installing the softener on your hot water line only to save money and benefit the environment. Softening hot water only means less water passes through the softener, which means it needs regeneration less often and less salt is used and recycled into the waste stream. Also, if cold water isn't softened, a cold glass of tap water still has the taste and health benefits of the water's mineral content. Softened water used for outside irrigation is also not a prudent use of resources.

Softening only the hot water does have a couple of disadvantages. First, if you wash your clothes in cold water, you won't get the benefit of soft water; but you can buy softening products to add to your wash. Second, and more importantly, if your water is very hard, the water will still be fairly hard when you mix hot and cold water together, and you will see only minimal benefit from the softener. You can also consider leaving the cold water tap in the kitchen unsoftened to use the water with beneficial calcium and magnesium nutrients for drinking and cooking, rather than the water with the sodium in it.

117. Does softened drinking water have any negative health effects?

The amount of sodium or potassium in the water after softening is relatively insignificant when compared with the amount found in many foods, so it should not be a problem unless you are on a very sodium-restricted diet. Contrary to some urban legends, drinking softened water does not drain calcium and magnesium from the body.

118. I have a water softener, but I still get spots on my bathroom tile and dishware. Why?

All water contains dissolved nontoxic minerals. Calcium, magnesium, sodium, sulfate, chloride, and bicarbonate are the most common. A water softener exchanges calcium and magnesium for sodium or potassium, so the water leaving the softener contains no calcium and magnesium (thus, no hardness) but more sodium or potassium.

Thus, all minerals are not eliminated during softening, just traded for other minerals. If you put the softened water in a dish to completely evaporate, the white stuff left over—although it would be different—would look the same and equal to the same amount as before the softener was installed.

119. How can I get rid of the precipitate deposits on my coffeepot or teapot and showerhead?

To remove a buildup of calcium and magnesium and other deposited minerals, fill your coffee pot with vinegar and let it sit overnight. Soak the showerhead overnight in a plastic bowl filled with vinegar. When you are done, carefully discard the contents of the plastic bowl down a drain and flush the container and sink drain with plenty of water. Rinse the coffee pot or showerhead thoroughly after treatment and before use. After the deposits are gone, help keep the buildup to a minimum by pouring the excess hot liquid out of your coffee pot when you are finished using it and wipe off the showerhead with a soft cloth.

White spots on glass shower doors are difficult to remove with vinegar because the spots dissolve very slowly. Commercial products are available, but handle these carefully and follow the manufacturer's directions. To prevent spots from forming, wipe the glass door with a damp sponge, squeegee, or towel after each shower. Home water that has been softened will likely produce fewer spots as well as spots that are easier to remove.

120. Some of my clear glassware comes out of the dishwasher with a rainbow sheen on it. What causes this?

Too much dishwasher detergent combined with soft water can leave an oily looking sheen on glassware. Also, some detergents can actually etch glasses with high silica content, resulting in a fine pattern of etching that appears as a rainbow coating.

BOTTLED WATER

121. Should I buy bottled water?

If your tap water meets all of the federal, state, or provincial drinking water standards, you don't need to buy bottled water for health reasons.

Many people turn to bottled water as a convenience, and in most cases, both bottled water and tap water are safe, healthy choices. But, if you want water with a different taste or without health additives such as chlorine, you can buy bottled water.

Bottled water is popular; US residents spend about $46 billion annually to buy this product, more than they spend on carbonated soft drinks. Recent surveys show that almost 20% of people in the United States say they get most of their drinking water from bottled water. Of course, in emergencies, bottled water or stored tap water can be a vital source of drinking water for people without water.

Cost and environmental protection are two major reasons to carefully evaluate choosing bottled water. Bottled water can cost up to 1,000 times more than municipal drinking water, which is about 0.30 cents per gallon compared with up to $5 per gallon for bottled water. On the environmental side, 25 million tons of plastic are used each year to bottle water, and much of that ends up in landfills or degrades to microplastics, which are of growing concern in the environment. Along with the energy used to produce and recycle the plastic, energy is used to transport the product to stores around the world. Although the International Bottled Water Association (IBWA) says that bottlers are protective of water resources, some manufacturers may be pumping up to 500 gallons per hour from valuable aquifers and other water sources.

Remember, if you use bottled water, consider it a food, refrigerate it after opening, and store it away from direct sunlight and other chemicals. Some plastics can break down in the sunlight, which causes chemicals to leach into the water.

122. Is bottled water regulated?

In both the United States and Canada, bottled water is regulated as a food product, which means it is regulated at the production source but not once the product has gone into distribution channels. The FDA is the US regulator, the Canadian Food Inspection Agency (CFIA) monitors bottlers under the Food and Drugs Act, and Health Canada establishes health and safety standards for bottled water and packaged ice and develops labeling policies related to health and nutrition. Provinces and states may regulate the bottled water sources, including drilling and construction practices, allowable rates of production, and watershed protection.

The FDA has established limits for more than 75 microbiological, physical, chemical, and radiological substances for both the source water and the finished bottled water product, and these standards must be as stringent as the USEPA standards for public water supplies. Bottlers are subject to annual inspections and must test weekly for the presence of bacteria in their water. Individual states and industry trade groups may impose additional regulations and requirements on bottlers. Regulations also require manufacturers' labels to list the contents and source of the bottled water.

In Canada, bottled water companies must adhere to similar quality standards, good manufacturing practices, and labeling requirements.

123. What do the labels on bottled water mean?

The label identifies the source of the water, such as spring, artesian well, or municipal water supply, and the type of water, such as mineral, purified, or sparkling. The FDA also requires nutritional information to be listed on the label of any food product, which includes total calories, calories from fat, sugars, protein, and fiber. Except for designer water or enhanced "sports" or "vitamin" waters, these elements are not found in most bottled waters. If the water contains more than an acceptable amount of a regulated substance, this must be stated on the label (e.g., "This product contains exces-

sive amounts of iron"). Types and sources of water are described as follows.

Artesian water or artesian well water is water that comes from a well drilled into a confined aquifer in which the water flows freely to the top of the well under natural water pressure.

Groundwater is water from an underground saturated zone—an aquifer—that is under a pressure equal to or greater than atmospheric pressure. Groundwater used for drinking water that is under the direct influence of surface water must be treated.

Well water is water from a hole—bored, drilled, or otherwise constructed in the ground—that taps the water of an aquifer.

Spring water is collected from an underground formation from which water flows naturally to the surface of the earth. Water must be collected at the spring or through a borehole tapping the underground formation feeding the spring. A natural force must cause the water to flow to the surface, and the location of the spring must be identified.

Mineral water is water that contains 250 mg/L or more of total dissolved solids, which is determined by evaporating the water and weighing the residue. Mineral water comes from a source tapped at one or more boreholes or springs and originates from a geologically and physically protected underground water source. Some mineral waters are naturally carbonated.

In Canada, bottled water classified as mineral or spring water cannot have its composition modified using any chemicals except for the addition of fluoride, carbon dioxide, or ozone.

Purified water or demineralized water is water from any source that is physically processed by distillation, deionization, reverse osmosis, or another process to remove impurities. Alternatively, water may be called deionized water if the water has been processed by deionization; distilled water if it has been processed by distillation;

reverse osmosis water if the water has been processed by reverse osmosis; and so forth. Sometimes, water that has been treated to this level must be remineralized to regain a pleasant taste.

Sparkling water is water that is enhanced with gaseous carbon dioxide—it's carbonated and contains small bubbles and an effervescent texture. Seltzer water and club soda are also carbonated, but they are considered soft drinks.

Sterile water or sterilized water is water that meets the requirements of the sterility tests in the *United States Pharmacopeia*.

The FDA requires the following additional labeling requirements:

- If the total dissolved solids content of mineral water is less than 500 mg/L or greater than 1,500 mg/L, the statement "low mineral content" or "high mineral content" must be added to the label, respectively.
- If the source of the bottled water is a community water supply or a municipal source, the source must be on the label, unless the water has gone through further treatment to render it as "purified" water.
- If the product states or implies that it is to be used for feeding infants and the product is not commercially sterile, "not sterile" must be added to the label.
- If the product does not meet the standards discussed previously, the terms "contains excessive bacteria," "excessively turbid," "abnormal color," "abnormal odor," "contains excessive chemical substances," or "contains excessive radioactivity" must be applied to the label, as appropriate.

Check the label carefully on any bottle of water to determine the sodium content, regardless of the general labeling. Some bottles labeled "sodium-free" do contain some sodium, which may be of concern for those on a sodium-restricted diet.

124. Is bottled water OK to store?

Yes, to a limit. Bottled water is a good source of drinking water during emergencies and when a person is on the go, but it does not store well because bottlers often remove or do not add any residual chlorine to the water, which would provide some protection against microbes over time. Unless stated otherwise, bottled water is not sterile. Ozone and ultraviolet light are the preferred disinfectant treatments used by bottled water manufacturers. Reverse osmosis, distillation, activated carbon, cation exchange, and microfiltration are also used. Chlorine disinfectants are not typically used, so microbes can grow in the water over time. These microbes are unlikely to make you sick, but to maintain freshness, the IBWA recommends that water bottles be labeled with the bottling date and be replaced every six months. Bottled water has a designated shelf life in a store, as well.

DOWN THE DRAIN

125. Where does water go when it goes down the drain?

If you are served by a public sewer system, all of the drains in your house are connected to a single pipe that leads to a main collection pipe under the street, where the wastewater is collected from all the homes in your area and is then transported to a larger pipe that collects water from other streets. The wastewater then flows into still-bigger pipes that connect various neighborhoods.

The pipes in the wastewater collection system become larger and contain more liquid the nearer they are to the wastewater treatment plant. Here, the wastewater is treated and cleaned so that it can be put back into the environment without harming anything and so is naturally recycled.

If you are not connected to a public sewer system, the liquid wastes from your home usually go into a septic tank, where most of the solids settle out. The water then overflows to a leach field, where pipes with holes in the bottom are buried in the ground. The water seeps out of these holes and into the ground, where it is naturally recycled.

126. What household chemicals can I safely pour down the sink or into the toilet?

Prioritize buying environmentally friendly products whenever possible so the ultimate disposal of the products does not become a problem. Next, try to buy just what you need so you won't have any or much left to dispose of. Finally, check with your local solid waste department or similar agency to learn of local rules and hazardous waste collection days. In Canada, local sewer-use ordinances control disposal in most municipalities.

If your home is served by a public wastewater system, these liquids can safely be poured down a drain at a time when you have been using a lot of water (such as washing clothes) to help dilute the wastes:

- Hand and laundry soaps
- Ammonia-based cleaners
- Drain cleaners
- Window cleaners
- Alcohol-based lotions and perfumes
- Bathroom cleaners
- Depilatories
- Hair relaxers
- Toilet bowl cleaners
- Tub and tile cleaners
- Water-based glues

After disposal, be sure to rinse the empty container with water several times. Of course, the safest course of action is not to put anything in your sink or toilet. Other wastes should be disposed of by following your local municipality's hazardous waste collection process.

127. How do I safely dispose of pharmaceuticals?

Don't flush pharmaceuticals down the toilet or pour them down the drain because wastewater treatment plants and septic systems are not designed to remove pharmaceutical waste, and the drugs often end up in our waterways.

Where available, take the medications to a hazardous waste collection site or take-back program at a medical care facility or pharmacy. Before taking any controlled substance to a collection event, however, check with the organizers to find out if they are authorized to accept the material.

The US Drug Enforcement Administration (DEA) has an interactive tool, "Everyday Takeback Day," on its website (www.dea.gov) that can direct you to a collection location based on your zip code.

More information about disposing of unused or expired pharmaceuticals is on the FDA's website, www.fda.gov.

128. I have a septic tank. Should I take any special precautions?

If you have a septic tank, check with your local authorities for any regulations or special conditions you should be aware of, but here are some general rules. First, remember that any substance you put down the drain into a septic system will likely seep into the local groundwater. Do not use any chemicals to clean your system; they may harm the system or the groundwater. Second, don't bother with septic tank additives or the addition of yeast because these elements really don't help the septic tank very much.

Third, minimize water usage. Don't run water continuously while rinsing dishes or thawing frozen food products (these are good conservation measures in any household). Consider limiting toilet flushes or putting a plastic bottle full of water in the toilet tank to reduce the amount of water used in each flush. Run only full loads when using a dishwasher or washing machine. Try to use the washing machine at times when water is not being used for other purposes. The reason for all of this is to pace the flow of waste through your septic system, which allows it time to do its job effectively.

Fourth, do not dispose of these items down the drain or toilet:

- Fats, grease or cooking oil
- Coffee grounds, meat bones, or other food products that don't biodegrade easily, even if they have gone through the garbage disposal
- Household cleaning fluids
- Automotive fluids such as gas, oil, transmission or brake fluid, grease, or antifreeze
- Pesticides, herbicides, or other potentially toxic substances
- Nonbiodegradable substances or objects such as cigarette butts, disposable diapers, and feminine hygiene products.

Finally, do not connect any footing or foundation sump pumps to the septic tank system. In all cases, follow your municipal or township codes and regulations.

FISH AND PLANT LIFE

129. How should I fill my fish aquarium?

First, allow at least 1 gallon (4 liters) of cold water to run from the tap before using the water to fill the aquarium to flush any copper or zinc from piping in your home. Tropical fish are very sensitive to small amounts of copper or zinc in their water. Then, hold a plate above the aquarium and pour the water onto the plate from about 1 foot (30 centimeters) above it before it hits the plate and flows into the tank. This adds air (oxygen) to the water. Let the water sit in the aquarium for an hour or two until it reaches room temperature.

If your water comes from a municipal supply that is disinfected with chlorine or chloramines, consult your local pet store to learn how to test for and remove the disinfectant in the water. Remove the disinfectant from the water in the aquarium before adding the fish, either by letting it sit for a few hours or by using a disinfectant removal product from your pet store. For tropical fish kept in a seawater environment, maintain the appropriate salt level using a saltwater additive and check concentration using a specific gravity indicator (i.e., a hydrometer).

130. I have trouble keeping fish alive in my fishpond. Is there anything I can do?

Domesticated fish die for many reasons. One problem is that waste products from the fish can decay and release ammonia, which is quite toxic to the fish and other aquatic life. Commercial products are available that will capture the ammonia and other harmful chemicals such as nitrate and nitrite and maintain the appropriate pH level. If you use a biological filter, the ammonia will be changed to a nontoxic chemical by microbial action. If you fill the pond with tap water, you may want to follow procedures similar to those used in filling an aquarium. Disease is also a possibility. Check with your local pet store for more advice.

If your fish are disappearing, it could be that raccoons or fishing birds, such as herons, have stolen them.

131. When I try to root a plant or grow flowers from a bulb in my house, the water looks terrible after a while. What will prevent that?

Put 1–2 tablespoons of activated carbon in the bottom of the bowl. This will help keep the water in better condition. Of course, changing the water frequently will help also.

132. Roses, azaleas, camellias, and rhododendron all require acid conditions. How should I adjust the acid content of my plant water?

The acid content of water is measured by the pH. Any number less than 7.0 indicates that the water is acidic, and pH levels greater than 7.0 are alkaline. Some plants like water with a pH somewhere between 6.5 and 6.8, lower than found in most tap waters. Acidic fertilizers, vinegar, or lime juice can be used to lower the pH of tap water before pouring it on these sensitive plants. Baking soda can be used to raise the pH of the water. Before deciding how much to add to a gallon of water, do some testing using a pH test kit that can be obtained from garden or pool supply stores. Also, check with the water supplier as to the pH and alkalinity of the tap water. The higher the alkalinity, the more difficult it will be to change the pH. Every state has at least one Cooperative Extension office at a state university that can provide detailed information on adjusting water quality to meet plant needs, and it can also test your soil and tell you what to add to it for a specific plant. You can find your extension office on the website www.extension.org.

133. I live in a very hard-water area, and I have a water softener. My plants don't seem to like my tap water. What can I do?

Water softeners usually replace the hardness chemicals (calcium and magnesium) with sodium. If you soften very hard water, you will wind up with a fair amount of sodium in your tap water, and some plants don't like sodium. Discuss this with your local garden store, or try one of these alternatives:

- Use reverse osmosis-treated water or distilled water for watering plants.
- Change your softener regeneration chemical from sodium chloride to potassium chloride (although this may be more expensive and harder to get).
- Collect water for the plants from a tap before the water gets to the softener, for example, from an outside garden hose connection.
- If you are now using sodium-softened water on your plants, water heavily to rinse off previously deposited minerals and discard excess water. Heavy sodium salt concentrations in the absence of calcium and magnesium may cause the soil to swell a bit and slow the growth of plant roots.

134. What causes the whitish layer on the soil of my potted plants?

Drinking water can contain many nontoxic chemicals and minerals. When the water on your plants evaporates, these chemicals—no longer dissolved—are left behind as a whitish layer. Using distilled water on your plants will avoid this problem. You can also catch rainwater for watering plants, but cover any water that is kept outside to prevent insect larvae infestation. If you have a dehumidifier, the water that comes out of it is good for watering plants. Distilled water, rainwater, and dehumidifier water will wash minerals out of the soil, however, so use a slow-release balanced fertilizer to replace these essential elements.

COSTS

135. What is the cost of the water I use in my home?

Most people pay for water delivered to their home according to the amount they use. In the United States, the water rate is usually based on each 1,000 gallons used; in most other industrialized countries, the charge is for each cubic meter (m^3) used. This is the "volume" charge for actual water used. Prices vary greatly, but a typical cost is in the United States is about $3.00 for 1,000 gallons (3.8 m^3). In Canada, where households tend to use less water, typical rates are about $1.00 per m^3 (264 gallons) in Canadian dollars, so if you do the math, rates are about the same in both countries.

Most utilities add a base charge of $5–10 per month for fixed utility costs in the distribution system (the cost to maintain the infrastructure to get the water to your home). In some places, this is called a meter charge. Still, either a gallon or a liter of tap water costs less than one penny.

Of the amount charged for 1,000 gallons, about $0.45–0.75 is for treatment; the rest is for paying the capital costs of the treatment plant and maintaining the pipes in the street, as well as the salaries of the employees who work for the drinking water utility and other fixed costs. If a waste disposal charge based on the water used is included in a water bill, some consumers have gone to the trouble and expense of having a second water meter installed to measure the amount of tap water used for lawns and gardens. There will be no wastewater disposal charge on the volume of this water used.

You can calculate the cost of water in your area by looking at your water bill and dividing the total cost for water by the total amount of water used (just use the water part of the bill if other costs are included).

136. How does the water utility know how much water I use in my home?

Most buildings are equipped with a water meter that measures the amount of water used. Some community water systems—particularly in smaller communities—are not metered, so households are charged the same flat rate each month. Financial experts consider this a less equitable and less conservative rate system. In communities with water meters, meters are read on a regular schedule, either in person or electronically from a remote location or passing vehicle. The previous reading is subtracted from the current reading to determine the amount of water used.

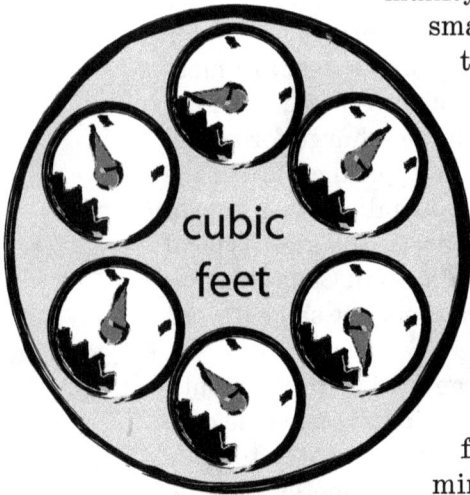

cubic
feet

137. How does the water company know that my water meter is correct?

Most water companies routinely test and periodically replace water meters on a rotating basis to ensure that the readings are accurate. If you notice that your recorded water use changes suddenly for no obvious reason, such as from having more people in your home, an extended trip away from home, or heavy lawn watering, report this to your water supplier so it can be investigated. Sometimes an undetected leak in the home can increase your water bill unexpectedly. In most instances, when a water meter is wrong, it reads less than what has been used. As a good citizen, you should report this to your water supplier just as you would when you think your meter might be reading too high.

138. We had a conservation drive in our area, and everyone cooperated. Then our water rates went up. Why?

Water suppliers have fixed costs—salaries, hydrant maintenance, upgrades to meet demand and regulatory requirements, mortgages, and so forth. They must collect this money regardless of water use, so when water volume goes down because of conservation by the public, the cost of each gallon of water used sometimes is raised to provide the water supplier with the money it needs to maintain its system and meet regulatory requirements. However, water conservation can eventually lead to stable rates for a longer period, because capital improvements or new investments can be postponed if demand does not increase.

Chapter 4

Sources

We forget that the water cycle and the life cycle are one.
—Attributed to Jacques Yves Cousteau

139. Where does my drinking water come from?

Drinking water comes from two major sources: water that flows above ground, known as *surface water,* and water that is pumped from beneath the ground, called *groundwater.* Surface water comes from lakes, reservoirs, rivers, and in a few cases, the ocean. Groundwater comes from wells that water suppliers drill into aquifers—underground geologic formations of permeable rock. Wells less than 100 feet (30 meters) deep are considered shallow wells, whereas deep wells can extend 1,500–2,000 feet (450–600 meters) below ground.

Springs are another source of freshwater. Springs begin underground as groundwater. When the water is pushed to the surface and flows out of the ground naturally, it becomes a spring. The water then may flow over the surface of the ground as a creek or river or may form a lake.

Most people in the United States—66%—live in areas served by large water systems that rely on surface water. The larger Canadian cities, including Montreal, Toronto, Edmonton, and Vancouver, also use surface water.

However, 80% of public water systems (PWSs) are in smaller communities that rely on groundwater. About 115 million people in the United States use groundwater supplied through individual wells and municipal groundwater systems. In Canada, about 2 million people are supplied by municipal groundwater systems.

If you get your water from a
PWS, your local water utility
details the specific source
of your drinking water
in its water quality re-
port to customers (called
a consumer confidence
report [CCR]) that is provided once or twice a year.
This information is also usu-
ally available on your utility's website.

140. How does nature recycle water?

Through the processes of evaporation, condensation, precipitation, and infiltration, known as the water cycle, the
total amount of water on the globe remains constant. Water
from oceans, lakes, rivers, ponds, puddles, and other water surfaces evaporates to become clouds. The clouds make
rain, snow, or sleet that falls to Earth to create rivers and
streams or seeps into the ground to form groundwater. All
this water flows to the ocean to start the cycle over again.

141. I have heard the term "mining groundwater." Is that anything like mining coal?

Yes, mining water is similar to mining coal. Just as when
coal is mined, less and less remains in the ground. Although groundwater is replenished by rainfall soaking
into the ground, the process is slow. When groundwater is
pumped out of the ground faster than it can be restored,
the groundwater is being mined. When this happens, the
underground level of the groundwater falls, and wells must
be drilled deeper to reach it.

142. Can groundwater be restored in any other way than the natural water cycle?

Yes. In recent years, some water suppliers have adopted a system called *groundwater recharge*. This is often performed in large basins where volumes of water spread over the land seep back into the aquifer. The water can be ultra-treated wastewater, diverted river water, surface runoff, or other excess water. This reused or excess water can also be returned to the aquifer through direct injection into recharge wells. On Long Island, New York, some groundwater used for air-conditioning is returned to aquifers through recharge wells. In Arizona, unused Colorado River water that the state owns and is delivered through a series of aqueducts is recharged into the ground for later use. This is called water banking.

Some water utilities have also adopted aquifer storage and recovery programs, which use similar recharge methods to inject treated wastewater into aquifers. In Virginia, Hampton Roads Sanitation District (HRSD) treats some of its wastewater effluent to drinking water standards and injects it back into the aquifer. By 2030, HRSD plans to be able to return 100 million gallons a day of treated wastewater to the aquifer.

143. How much drinking water is produced in the entire United States each day, and how much is used for industrial purposes and irrigation of crops?

In 2015, total water use in the United States was estimated at 322 billion gallons (1.2 trillion liters) of water daily. Public and private water supplies accounted for about 40% of that total, and the rest was used for irrigation, livestock, aquaculture, industrial, mining, and thermoelectric power generation. Irrigation and power generation use the most water. Daily irrigation use can be significant, depending on the location and the time of year. It takes about 50 glasses of water just to grow enough oranges to produce one glass of orange juice. One estimate puts the total amount used for irrigation at 118 billion gallons (446 billion liters) a day. Of course, crop irrigation water is diverted from the source and does not get treated as tap water does.

144. How much of the Earth is covered with water, and how much of that is drinkable?

Close to three-quarters of Earth's surface is covered with water, but less than 1% of that water is suitable and available for drinking water using conventional water treatment.

Of all the water in the world, 97% is in the oceans, and about 1.7% is locked up in ice and snow; another 0.3% of freshwater is in aquifers that are too deep to access, and about half of the 0.3% of groundwater that is accessible is too salty to use efficiently.

145. What country has the most potable water?

If money were made out of water, Finland would be the richest country in the world, according to the Water Poverty Index compiled in 2002 by Keele University in the United Kingdom. This ranking measures water availability in a given country and people's capacity to access that water and includes the reliability or variability of the resource. Because of its plentiful supply of surface water and groundwater, an infrastructure that provides ready access, and several other factors, Finland ranked richest among 147 countries studied, and Canada ranked second. Haiti ranked last. Penalized for what a panel of water experts judges saw as its wasteful use of water, the United States ranked 32nd.

A different study in 2020 that focused on health outcomes ranked the United States 23rd worldwide in drinking water health safety.

146. Can ocean water be treated to make drinking water?

Ocean water contains so much salt that at least 99.2% of the salt must be removed to avoid a salty taste in drinking water, but it can be treated to make drinking water through a process called *desalination*. Many people around the world, particularly in arid countries in the Middle East and Africa, as well as in coastal communities in North America, rely on this process. However, the process is more expensive than traditional treatment because it requires a lot of energy. The cost of desalination has been estimated at $2–6 for each 1,000 gallons ($0.53–1.59 per 1,000 liters) instead of the average of $0.57 per 1,000 gallons ($0.15 per 1,000 liters) for conventional treatment. Of course, the cost of ocean water treatment must be compared with the total cost of providing water from another source, which includes building dams and pipelines, installing pumps, and so forth.

147. I've heard about towing icebergs to areas that are short of water as a source for drinking water. Would that really work?

Whenever we are faced with water shortages, lots of obscure ideas are floated for consideration. This is one such idea that has been considered in various forms since the 1970s. It's never been done commercially because it would be so expensive and difficult to manage. It could theoretically work, although melting would take place during the voyage. Even though icebergs are floating in saltwater, the ice has very little salt in it—it's basically compressed snow. A melted iceberg could provide potable water, but it would still require treatment.

148. Why are rivers dammed to create reservoirs? Can't the utility just draw what it needs from the flowing river or underground aquifers? What about the environmental impacts?

Water systems need to have enough water stored to meet the demands of its users, including residential, industrial, commercial, and public use, such as fire control. If a river is running low—say, in the fall or during a drought—the water in a reservoir can be drawn on to ensure adequate supplies. Also, it's easier for a water treatment plant to treat the reservoir water than river water because flowing river water can vary quite a bit in water quality, whereas water stored in a reservoir tends to have consistent quality over time.

Environmental impact assessments must be conducted for any new dams or reservoirs. This process includes evaluating the effects on wildlife and their habitats according to the Endangered Species Act and other regulations and assessing instream flows of affected rivers to ensure adequate water for existing aquatic life.

149. Can wastewater be treated to make it into drinking water?

Yes, and this practice, which is commonly called *reuse*, is gaining more and more acceptance around the world. The water industry sometimes calls this *reclaimed* water, and most often, the water is used for nonpotable purposes such as park or golf course irrigation, which saves more freshwater for drinking and bathing. Wastewater is already thoroughly cleaned before it is discharged to the environment, and when it is reclaimed for reuse, it is cleaned again. When it is treated for drinking water, it is cleaned again and again.

If you think about it, most communities that draw their drinking water from rivers or streams already are receiving reused water from an upstream community that discharges its treated wastewater into the same body of water. And, of course, nature reuses water through the water cycle.

150. Is climate change affecting our water supply?

Water industry researchers tell us that climate change has significant implications for water supply management because increasing temperatures will increase evaporation rates, change precipitation patterns, affect runoff and water quality, and change water-use demands.

Climate change has already had some effect on our water supply. Salinity is increasing as seawater replaces freshwater as more and more groundwater is pumped to meet demand. Earlier or reduced snowmelt in the mountains affects supplies downstream. Changing water temperatures can affect the amount of natural organic matter in the water, encourage algal growth, and subsequently affect water quality. Wildfires can affect water supplies during the active fire stage (as ash contaminates the water sources) and for years afterwards, as the absence of vegetation contributes to erosion that can flush large quantities of ash and sediment into streams, rivers, lakes, and reservoirs, making water difficult or impossible to treat.

Water professionals are learning how to adapt to the challenges of climate change with processes such as water banking and reclamation, but we can all help preserve our precious freshwater by conserving when we can and protecting the sources from pollution.

QUALITY

151. Which is more polluted, groundwater or surface water?

It depends on what you call pollution. Because surface water can be contaminated by municipal wastewater, industrial discharges, transportation accidents, and rainfall runoff, it contains many pollutants but not much of any one chemical. Groundwater, on the other hand, may contain pollutants such as arsenic, nitrates, radioactive materials, and comparatively high amounts of a few organic chemicals such as cleaning fluid or gasoline. Therefore, both may be polluted but in different ways.

Another difference is that the degree of pollution may change rapidly in surface waters, whereas pollution levels change very slowly in groundwater. Your water supplier can tell you what contaminants it has found in your source water, but the quality of treated water is tested more frequently than the quality of source water.

152. What is the major cause of pollution of surface waters used for drinking water?

In towns and cities, the major source of pollution is rainwater that flows into street catch basins (called urban runoff or stormwater runoff). Although this rainwater alone is not necessarily harmful, it can pick up untreated waste products from our streets and yards and then dump it directly into rivers, streams, and lakes that are drinking water sources. In rural areas, sediment that gets into the surface water can disrupt the natural biological processes in the water, making treatment more difficult.

153. What is the major cause of pollution of groundwater (well water) used for drinking water?

Groundwater can be contaminated by substances from the land surface that move through the soil and end up in the groundwater and from natural materials that are in contact with the water underground. Man-made influences include outflow from septic systems, leaking fuel tanks, pesticides and fertilizers applied to the surface, and improperly disposed-of chemicals. Natural materials that are considered pollutants in groundwater include arsenic, iron, manganese, and radon. These substances enter the water when it is in contact with rocks containing the minerals.

154. Is urban runoff treated before being discharged into drinking water sources?

Because of concerns about pollution, as mentioned in the previous question, more and more communities are capturing and treating their stormwater before it reaches drinking water sources. One way is to hold the water in detention ponds so some of the sediment and associated pollutants settle out before discharging it. Manufactured devices are also available to treat runoff, including catch-basin inserts that capture debris and filter out some pollutants, such as oil; hydrodynamic separators that remove pollutants from the water through filtration or other means; and underground storage units. Such practices are expensive and can treat only a portion of the runoff.

Some communities, especially older cities, have combined sewer systems (the sanitary sewer and the storm sewer share piping), so all water is routed to a wastewater treatment plant. As cities grow, however, the combined sewer systems become too small for all the water and can overflow during a heavy storm, releasing untreated or partially treated sewage to the environment. As a result, many cities are working on creating a separate storm sewer system that will capture and treat the water that flows into the storm sewer before it's discharged.

155. We used to hear a lot about acid rain. What is acid rain? Is it still a problem, and does it affect water supplies?

Acid rain forms when pollutants from natural sources, such as volcanoes and decaying vegetation, react with pollutants from man-made sources, primarily sulfur dioxide and nitrogen oxides. Most emissions come from burning fossil fuels such as coal to generate electrical power and gas-burning vehicles. The rain picks up these pollutants and falls on surface waters and land that runs into surface waters, increasing acidity in the water. Natural plant and aquatic life can have a hard time surviving in these conditions.

Because of the Clean Air Act and other environmental regulations, these types of emissions have declined in recent years. Acid rain is not the threat it used to be, but it does remain in some areas.

Scientists have predicted that further emission reductions are necessary to prevent increasing damage to North American ecosystems, particularly in New England, some parts of the Rocky Mountains, and much of Eastern Canada.

Water treatment plants can adjust the pH levels and balance the water between acid and alkaline, so it isn't a problem with the water you receive at the tap.

156. Do oil spills pollute drinking water sources?

Although oil spilled in the oceans is bad for the environment, it is not a danger to drinking water sources unless you live on the coast and your water utility uses a treatment process that includes desalinating seawater. Oil spills from ship and barge accidents on lakes and rivers do contaminate surface water sources, however. Many highways and railroad tracks and some pipelines pass over or near drinking water sources, which creates the potential for major contamination if an accident occurs. A motor vehicle accident or improper disposal of oil from your car can also cause oil pollution. Drinking water contaminated with even a little bit of oil has such a bad taste that most people regard it as undrinkable.

Groundwater sources would be affected by these types of accidents if they occurred in the recharge area (where water seeps downward to add water to the underground aquifer). Major oil spills in these areas can cause nearly irreversible damage to groundwater because groundwater moves so slowly.

157. I live downstream from a nuclear power plant. Should I worry about radioactivity in my drinking water?

It depends. Nuclear power plants operate under strict guidelines from the Nuclear Regulatory Commission (NRC), and most operate safely to prevent dangerous radioactivity from getting into the water. However, by the NRC's own admission, "a nuclear power plant may deviate from normal operation with a spill or leak of liquid material."[1] In particular, "several instances of abnormal releases of liquid tritium from several nuclear power plants ... resulted in groundwater contamination," according to a 2006 NRC report. However, tritium occurs naturally in the environment and the foods we eat and disperses quickly once it enters the body, and the health risk of drinking it at levels found in some of the contaminated groundwater is considered much less than that of the radiation exposure from many medical procedures.

Water systems monitor the radioactive content of source water to ensure that excessive amounts are eliminated before it becomes tap water.

---

1 NRC (US Nuclear Regulatory Commission). 2024. "Backgrounder on Tritium, Radiation Protection Limits, and Drinking Water Standards." www.nrc.gov/reading-rm/doc-collections/fact-sheets/tritium-radiation-fs. html (accessed June 11, 2025).

158. How can I help prevent pollution of drinking water sources?

A lot of water contamination occurs when people or companies improperly dump solvents, cleaners, oils, pharmaceuticals, and other chemicals in the ground, down storm sewers, or in septic and wastewater systems. Much of this pollution cannot be traced to a single source, so the US Environmental Protection Agency (USEPA) calls this non-point source pollution.

The USEPA recommends these actions by individuals:

- Keep litter, pet waste, leaves, and debris out of street gutters and storm drains.
- Apply lawn and garden chemicals sparingly and according to directions.
- Dispose of used oil, antifreeze, paints, and other household chemicals properly, not in storm sewers or drains. If your community does not already have a program for collecting household hazardous wastes, ask your local government to establish one.
- Clean up spilled brake fluid, oil, grease, and antifreeze. Do not hose them into the street where they can eventually reach local streams and lakes.
- Control soil erosion on your property by planting ground cover and stabilizing erosion-prone areas.
- Encourage local government officials to develop construction erosion/sediment control ordinances in your community.
- Have your septic system, including the leach field, inspected and pumped at least once every three to five years so that it operates properly.
- Purchase household detergents and cleaners that are low in phosphorus to reduce the amount of nutrients discharged into our lakes, streams, and coastal waters.

Remember your drain is an entrance to your wastewater disposal system and eventually to a drinking water source. Discharges from septic tank drain fields may pollute groundwater. Treat your wastewater system with respect.

More information can be found about nonpoint source pollution at these websites: the USEPA at www.epa.gov; The Groundwater Foundation at www.groundwater.org; and the Local Government Environmental Assistance Network at www.lgean.net.

159. I have a private well. Where can I get my water tested, and what should it be tested for?

Well owners should have their water tested annually for nitrates, total coliform bacteria, pH, and total dissolved solids. In addition, if the well or the home is new to the owner, the water should be tested for radon and arsenic. Well owners should also ask a local or state agency such as a health or agricultural department about any specific contaminants of concern in their local area (e.g., pesticides or fertilizers) that should be included in the testing. Most state health departments have programs for private well owners, and they are your best source for information on testing and local laboratories.

The USEPA's website has a lot of good information for private well owners at www.epa.gov, including information about how to protect your well, links to public agencies by region, and other materials. The Groundwater Foundation also has a free information service for well owners called wellcare Well Owners Network; information is available at www.watersystemscouncil.org.

160. How can I protect my private well water supply?

You can protect a private water supply by carefully managing activities near the water source, including following these recommendations from the USEPA.

- Have your septic tank pumped and system inspected regularly, at least every one to three years.
- Use fertilizers sparingly, and sweep up driveways, sidewalks, and roads.
- Never dump anything down storm drains or onto the ground around the well, house, or underground piping.
- Revegetate or mulch disturbed soil as soon as possible.
- Clean up spills of vehicle fluids or household chemicals and properly dispose of cleanup materials.
- Minimize pesticide use and learn about integrated pest management.
- Direct roof drains away from paved surfaces and bare soil.
- Take your car to a car wash instead of washing it in the driveway.
- Check your car for leaks and recycle motor oil.
- Pick up after your pet.

Chapter 5

Distribution

When you put your hand in a flowing stream, you touch the last that has gone before and the first of what is still to come.
—Leonardo da Vinci

161. Are the pipes that carry drinking water from the treatment plant to my home clean?

A well-run public water system will have a program of flushing and cleaning the distribution pipes to ensure that the water delivered by the distribution system is free from impurities. Water distribution pipes made of metal or concrete can corrode internally because of corrosive water quality characteristics that react with the inner surface of the pipe. Most corrosion byproducts are not harmful to human health, but they can cause nuisance issues, such as spotting on laundry or a metallic taste in the water.

From the street to your home, the pipe carrying the water is called a service line, and some of these service lines are made of lead. Lead is a neurotoxin and is particularly harmful to children younger than six years. Water utilities condition the water to prevent corrosion of the lead from service lines, as well as the iron in the water mains that convey water to the service lines. The Lead and Copper Rule Improvements (LCRI) require the identification and removal of lead-bearing materials in the distribution system, including lead service lines, and federal funding has been set aside to help utilities pay for this effort.

If chemical reactions in the water cause sulfates and carbonates to form, the sulfates and carbonates can stick to the walls of the pipe and form scale. Scale can clog fixtures and reduce the flow of water in pipes. Periodic water pipe flushing keeps the buildup of corrosion byproducts and scale to a minimum. How often a water system needs flushing depends on water chemistry and pipe material.

To flush the pipes, valves are closed and fire hydrants opened, forcing water pressure to concentrate in certain pipes. Then, the water rushes out of a hydrant, along with the buildup from the pipes. Alternatively, pipes may be cleaned by using water pressure to force a tight-fitting plastic device called a "pig" through the pipe. The pig scrapes the pipe walls clean. Some utilities use an icy slurry instead of the plastic pig and it works the same way. Corrosion byproducts and scale are then flushed out of the system through an open fire hydrant. Similar pipe-cleaning devices with other designs are also available.

Keeping water pipes clean is a big job, as there are more than 2.3 million miles (3.4 million kilometers) of pipes in the United States and Canada.

162. I have seen work crews cleaning water mains, and the water they flush out looks terrible. How can the water be safe if the pipes are so dirty?

As discussed in the previous answer, water suppliers' regular flushing and cleaning programs remove corrosion and scale from the walls of several miles of pipe. When this material flows out of a fire hydrant all at once, it looks worse than it really is. If you watch the workers do this, you will notice that the water clears up rather quickly. Plus, the freshwater that is being used to flush the pipes contains a disinfectant, and this clean water is what replaces the dirty water in the pipes.

163. What are distribution pipes made from?

Pipe may be made of ductile iron, steel, concrete, polyvinyl chloride (PVC), or high-density polyethylene (HDPE). Ductile-iron, steel, and concrete pipes are strong and durable and can be used in systems with very high pressures. PVC pipe is light, inert to most chemicals, and noncorrosive. It is easy to install and can be cut with a handsaw. HDPE pipe is softer and more bendable than PVC and is suitable for lower pressures and applications with bends in the pipe. Lead was used in some areas in service lines, which is the pipe going from the water main in the street to your home, but it is no longer permitted. Water utilities are required to treat the water so it doesn't corrode the lead pipes and are required to implement a program to replace lead pipes, but if you have any concern, you should contact your water supplier.

164. Why are fire hydrants sometimes called fire plugs?

Long ago, tap water was distributed in hollowed-out log pipes. When water was needed to fight a fire, the pipe was dug up and a hole was drilled in the wooden pipe. When the fire was over, a wooden plug was used to close the hole. These fire plugs were then marked for possible future use.

165. We don't use much water for drinking. Why does all the rest need to be treated so extensively? This seems unnecessarily expensive.

Actually, the cost of water treatment itself is a small part of your water bill, but that is not the real answer. An additional distribution system could be developed in which two pipes come from the treatment plant, through the streets, and into your home. This dual distribution system would feature a small pipe for drinking water and a larger pipe for all of the other water uses (toilet flushing, lawn watering, and so forth). To install a dual system in existing cities would cost a lot and would disrupt services during the conversion. However, many communities are now recognizing the value of *reclaimed* or *reused* water as a way to satisfy water demands, particularly for uses that do not require potable-quality water, and are starting to install systems in new construction and to modify their existing systems to allow using reclaimed water for nonpotable uses. In any case, the water provided for purposes other than drinking still needs to be treated to ensure minimum health risk.

166. How does a water utility detect a major leak in the distribution piping system?

A major leak can be detected by

- visual observation (water on the ground or spraying water);
- a loss in water pressure;
- a depression in the ground or road over a water main;
- reports by public-minded citizens;
- sensitive listening devices that detect the sound of the leaking water underground; and
- a satellite that can detect underground chlorinated water, meaning a nonsurfacing leak can be detected from space.

Also, in metered systems, water utilities regularly compare the amount of water produced versus the amount that passes through customer meters. This method provides an excellent accounting of overall water loss.

Stopping leaks is important to water suppliers because leaks waste water, adding costs to both the water supplier and you. Water utilities don't get paid for the water that is lost to leaks, but the cost for treatment and distribution must still be passed along to customers. The national average for unaccounted water from all sources is 15% of treated water, although many suppliers keep such losses lower. Gas companies only lose about 5% of their product.

Any leakage that occurs within the boundary of your property after the water meter is your responsibility and must be repaired at your own expense. Prompt repair is to your benefit, because as long as a pipe is leaking, your water bill will be higher.

167. I've been hearing a lot about "aging infrastructure" of drinking water systems. Is this a real problem?

Most pipes in a municipal water distribution system were laid before many of us were born and are failing at different rates, depending on the type of pipe, ground conditions, usage, and other factors. Repairing the miles of aging pipes is a massive job, not to mention the cost. In fact, much of the drinking water infrastructure in the United States will need to be replaced in the next three decades. Paying for these replacements will be challenging.

The cost estimate to replace all the old drinking water pipes in the United States is more than $600 billion. The federal and state governments offer some relief to utilities to offset the cost of repairs, generally in the form of low-interest loans, but in the end, water rates will have to rise to pay for repairing and replacing current water infrastructure.

Federal funds are becoming more available for these programs, particularly for replacing those pipes made of lead.

168. Fixing a broken water pipe looks like a dirty job. How is the inside of the pipe cleaned afterward?

After work is done, the pipe is flushed to clear it of loose debris that may have collected during the repair. Then the pipe is filled with water containing a disinfectant—usually some type of chlorine. Holding this water in the pipe for a time kills bacteria and other organic contaminants in the system. This is followed by testing to be sure the water is safe.

This is not the end of the story, however. The utility must take care in disposing all this water that contains so much chlorine. State, provincial, and federal regulations control its disposal. A chemical must be added to react with the chlorine and inactivate it before the water can be flushed out of the pipe and discharged, unless the highly chlorinated water is discharged to a wastewater treatment plant or another location where it will not have an adverse effect on the environment.

169. What causes low water pressure?

Temporary low pressure may be caused by heavy water use in your area—lawn watering, a water main break, fighting a nearby fire, and so on. If the pressure drop is significant, a water supplier is required by regulation to issue a "boil-water notice or advisory" to its customers. This is because contamination can be "siphoned" or pulled into the pipes if the pressure is too low.

Permanent low pressure could be caused by the location of your home (on a hill or far from the pumping plant or water tank) or pipes that are too small. If you have an older home, the pipes in your home could have a lot of scale in them, leaving little room for the water to flow. Sometimes, water pressure problems are as simple as a plugged faucet aerator, and sometimes, the design of the interior plumbing can cause low pressure.

Low pressure is more than just a nuisance. The water utility depends on pressure to keep out contamination and to ensure enough flow to fight fires. If the pressure drops, the possibility of contaminants entering the drinking water increases. One of the causes of poor-quality water in some cases is low pressure in the distribution system that allows contamination to enter the pipes. You should report any unusual drop in water pressure to your water company.

Many areas have minimum standards for pressure. For example, 20 pounds per square inch (psi)—140 kilopascals (kPa)—when water use is at a maximum is a common standard. Most systems have pressures three to four times the minimum.

You can tell you may have low pressure if flows from your faucets at home are much lower than elsewhere in your area: at work, in a restaurant washroom, or in a friend's home elsewhere in the city, for example.

The only way to cure permanent low pressure is to have the supplier change the system, adding more pumps or bigger lines. If the problem is in your home (more of a nuisance than a potential health hazard), check the faucet aerators for any accumulated sediment, and discuss your options with a reputable plumber. For example, there may be a restriction in your piping, valves, or water meter that could be easily corrected. Or, if you have newer fixtures, they may be designed to restrict flow as a conservation measure.

You may be surprised to learn that you can also have too much pressure. Some homes need pressure regulators to avoid damaging household plumbing from very high water pressures.

170. What are cross-connections?

A cross-connection is a connection between a drinking water pipe and a contamination source. Here's a common example: You're planning to spray weed killer on your lawn. You hook up your hose to the outside spigot on your house and to the sprayer containing the weed killer. If the water pressure drops at the same time you turn on the hose, the chemical in the sprayer may be sucked back into your home's plumbing system through the hose. This is called *backsiphonage* and would seriously contaminate the water system in your home. If your hose was connected to a fire hydrant or a public access faucet (e.g., at a campground), then the weed killer would be sucked into the public water supply.

Backsiphonage can be prevented by using an attachment on your hose called a *backflow-prevention device.* The simplest backflow-prevention system is an air gap, which is a physical separation of the supply pipe by at least two pipe diameters vertically above the overflow rim of the receiving vessel (the sprayer containing the weed killer in the example). A hose bib vacuum breaker (a mechanical device that allows air into the piping system, placed where a hose is connected to the water line) installed on the outdoor spigot will also work.

More sophisticated backflow-prevention devices are mandatory for certain industrial and commercial operations, such as dry cleaners and restaurants. Backflow preventers are usually required by the water utility on irrigation systems at commercial properties. Most water suppliers have cross-connection control programs, particularly in major cities.

171. Why is some drinking water stored in large tanks high above the ground?

Aboveground storage ensures that water pressure is fairly constant in the distribution system and water volumes are sufficient to fight fires, even if the electricity that runs the water pumps is off. Elevated storage tanks make use of gravity to keep the water flowing. The storage tanks also provide a steady supply of drinking water when water use is high. The tanks are usually filled in the afternoon or late at night when

drinking water use is low. Water suppliers must pay attention to the operation and maintenance of storage tanks so the water circulates frequently and stays fresh.

172. I see on TV that asbestos is harmful to one's health. Aren't some of the water pipes in the street made of asbestos? Is that OK?

Yes, asbestos fibers are a health hazard when breathed in, and no, water pipes are not made of asbestos. Some are made of a product called "asbestos–cement." This type of pipe has two main advantages: It does not rust, so there are no red water problems; and it is very smooth inside, so water flows through it very smoothly. If the water is treated so that it is not acidic, the asbestos fibers are not released. Plastic pipe has largely replaced asbestos–cement pipe for new construction.

Chapter 6

Conservation

When the well's dry, we know the worth of water.
—Benjamin Franklin, *Poor Richard's Almanac,* 1746

173. What indoor home activity uses the most water?

Inside the home, we use water for consumption, washing ourselves and our clothes and dishes, but nearly 30% is flushed down the toilet. According to the US Environmental Protection Agency (USEPA), a typical household of four uses nearly 400 gallons of water each day. In addition to the 30% used for toilet flushing, clothes washing accounts for 26% , followed by showers at 20% and faucets (dishes, washing hands, brushing teeth, etc.) at 19%.

To put this in perspective, a person needs only 5 gallons of clean water a day to meet basic needs, according to the World Health Organization. Many people in undeveloped countries do not have access to even that much clean water.

174. How much water can we save by installing new fixtures?

Water-efficient toilets can save the average home more than 13,000 gallons (49,210 liters) each year, according to the USEPA. Low-water-use toilets require less than 1.3 gallons (4.9 liters) for each flush, compared with 7 gallons (26.4 liters) for pre-1980 vintage toilets and 3.5 gallons (13.2 liters) for more recent, less efficient models. If you replace your old clothes washer with a high-efficiency washing machine, you can reduce water consumption from

115

40 gallons for each load to between 7 and 15 gallons (26.4 and 56.8 liters) for each load. And, if 1 out of every 10 US homes upgraded to water-efficient fixtures, more than 120 billion gallons (454 billion liters) of water and more than $800 million would be saved annually. You can find out about water-saving appliances by consulting WaterSense, a USEPA program that provides information on products and programs that save water without sacrificing performance, online at www.epa.gov/watersense.

Using less water conserves in other ways as well. It takes a considerable amount of energy to treat and deliver the water you use every day, and the USEPA has determined that if just 1 out of every 100 US homes was retrofitted with water-efficient fixtures, about 100 million kilowatts per hour of electricity could be saved per year— avoiding 80,000 tons (72,600 metric tons) of greenhouse gas emissions. That is equivalent to removing nearly 15,000 automobiles from the road for one year! And it will help your electric bill, too.

175. Why can't I just put a brick in my toilet instead of replacing it?

Putting something in the toilet tank that takes up space means that less water is needed to refill the tank after a flush—and that is a good idea—but putting a brick in your toilet tank is not smart. A brick tends to crumble and might damage the toilet's flushing mechanism or clog the water inlets. Instead, use a plastic jug filled with water. Because some toilets require a certain volume of water to work right, be sure to test the toilet to make sure it's still flushing well after any changes.

176. Before I replace my showerhead, how do I measure how fast my shower is using water?

First, know that the water coming from a low-flow showerhead should not exceed 2.5 gallons (9.5 liters) per minute.

Next, get a bucket and a stopwatch or a watch with a second hand. Make sure the bucket has a 1-gallon (3.8-liter) mark on it. If it doesn't, add a gallon of water and mark the level. Set the shower flow just as you would when showering. Catch all the water in the bucket for 24 seconds. If after that time the water level is near the 1-gallon mark, your showerhead is flowing at the recommended amount. If the level is over the 1-gallon mark, replace your showerhead with a new low-flow model.

177. What about outside use—how can I conserve water there?

At least 30% of water used by a single-family suburban household is for outdoor irrigation, and much of that goes to waste through evaporation or runoff caused by overwatering. To conserve water, consider these landscaping tips:

Replace your fescue lawn with a native ground cover that doesn't require as much water as grass.

- Maintain a lawn height of 2½–3 inches to help protect the roots from heat stress and reduce the loss of moisture to evaporation.
- Promote deep root growth through proper watering, aeration, fertilization, grass-clipping control, and attention to lawn height. A lawn with deep roots requires less water and is more resistant to drought and disease.
- Avoid planting turf in areas that are difficult to irrigate properly, such as steep inclines and isolated strips along sidewalks and driveways.
- Aerate clay soil by adding compost periodically to help the soil retain moisture.
- Mulch around plants, bushes, and trees to help the soil retain moisture, discourage the growth of weeds, and provide essential nutrients.
- Plant in the spring or fall, when watering requirements are lower.

117

- When choosing plants, keep in mind that smaller ones require less water to become established.
- Contact your state's Cooperative Extension office to get information on how to conserve water through best management practices for your region. You can find that office by visiting www.extension.org.

For other outdoor uses, consider these tips:

- Use porous materials for walkways and patios to keep water in your yard and prevent wasteful runoff.
- Use an adjustable hose nozzle for plant irrigation and be sure to shut water off at the house connection each time the exterior faucet is used.
- Use a broom or rake to remove debris from driveways and walkways, not water.
- If you have a pool, keep the water level low to minimize splashing, and use a cover to slow evaporation. An average-sized pool can lose about 1,000 gallons (3,785 liters) of water every month if left uncovered.
- Use a bucket of soapy water and a hose with a shutoff nozzle to wash your car.

178. How should I irrigate my lawn and garden to avoid wasting water?

Water early in the morning to avoid excessive evaporation from sun and wind; water pressure is typically higher then, too. Water your lawn for extended periods a couple of times each week, rather than every day, to allow deep moisture penetration. An inch of water a week is a good rule of thumb, but this varies for different grasses and different parts of the country.

Before watering, use a trowel to check the root zone of your lawn or garden for moisture; if the soil is still moist 2 inches below the surface, you don't need to water. Or install an automatic irrigation sensor that measures the moisture content of the soil and controls when and where an area is watered. This is particularly helpful to prevent watering when it has just rained.

For a real difference, replace your sprinkler system with a drip-irrigation system. According to the USEPA, a drip system uses between 20 and 50% less water than conventional inground sprinkler systems, and no water is lost to wind, runoff, and evaporation.

As noted previously, every state has an extension office that offers solid, science-based information on how to best manage your outdoor plantings, including your fescue lawn. You can find your extension office by visiting the website www.extension.org.

179. Can I save water by changing my landscaping to xeriscape?

Xeriscape is a landscaping design that combines water conservation practices with creative landscape design that emphasizes the use of plants that are appropriate for your climate and soil conditions. Xeriscape calls for grouping plants according to their watering needs, along with proper soil preparation; using shade; rethinking traditional grass lawns; taking advantage of natural runoff; planting in low-irrigation areas; and using mulch. Xeriscape, combined with efficient irrigation, can save a lot of water and create a beautiful landscape.

180. I heard that some water conservation measures are required by law. Is this true?

In the United States, low-volume toilets and shower fixtures have been required since 1994 in new home construction, and since 1997 when fixtures are replaced in homes and in businesses, according to provisions established by the Energy Policy Act of 1992. This law has helped fund the toilet rebate programs that some water utilities offered. Some states, led by California, have adopted water-efficient landscaping regulations and other restrictions on water use.

In some cities or states, immediate and significant water restrictions can be put in place in times of emergency or serious drought.

Canada has made significant efforts to introduce requirements and programs encouraging water conservation through water-efficient appliances and fixtures. Several Canadian utilities have initiated low-flow toilet installation programs.

Limitations on water supply, as well as limitations in the capacity of utility infrastructures, will eventually require utilities to introduce water-saving measures, and, if the public doesn't willingly participate in conserving water, more regulations may be enacted to ensure there is enough of this vital resource to go around.

181. Which uses more water, a tub bath or a shower?

A full bathtub requires about 70 gallons (265 liters) of water, while taking a 5-minute shower under a low-flow showerhead uses 10–25 gallons (38–95 liters). If you do take a bath, don't run water down the drain while it heats up; adjust the temperature as you fill the tub.

If you have a shower in a tub instead of a stall, close the drain when you are showering, and you will see how much less water is used for showering compared with filling a tub.

182. I leave the water running while I brush my teeth. Does this waste much water?

You bet! Leaving the water running is a bad habit; about 4–6 gallons (20–25 liters) of water go down the drain needlessly every time you brush because the average bathroom faucet runs at a rate of about 2 gallons (7.5 liters) a minute. Turning off the water when you are not using it will save water and save you money.

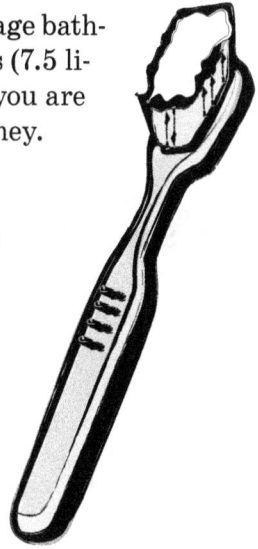

183. I use a lot of water in the kitchen. How can I conserve water there?

- Remove the residue from each cooking utensil and dish without using water, and don't rinse them before putting them in the dishwasher.
- Clean vegetables in a pan of water rather than under running tap water, and then use that water to give your plants a drink.
- Use the garbage disposal sparingly.
- Run the dishwasher only when it is full.
- Keep a container of water in the refrigerator instead of running the tap to cool your drinking water.
- Use a pail or basin instead of running water for household cleaning. A sponge mop will use less water than a string mop.
- Presoak grills or oven parts overnight when they need cleaning. Wash with an abrasive scrub brush or pad and use plenty of "elbow grease" to minimize water use.

184. Many water quality problems in the home—lead, red water, sand in the system, and so forth—are cured by flushing the system. Isn't that a waste of water?

Yes, but you can avoid losing this water by catching it in a container and using it for plant and garden watering. Flush water keeps lead or rust out of the water you consume, and that benefit will outweigh the loss of a few gallons. Try to use your flush water for nonpotable uses, but if you can't, don't feel too bad. This water has served a useful purpose.

185. My water faucet drips. Should I bother to fix it?

Yes. Drips waste precious water, and you will pay for it eventually. To find out how much water is wasted, place an 8-ounce (236-milliliter) measuring cup under the drip and watch how many minutes it takes to fill it up. Divide the filling time by 90 to get the gallons of water wasted each day. For example, if your faucet fills the measuring cup in less than 30 minutes, it could be dripping 60 times a minute (once each second), adding up to more than 3 gallons (12 liters) each day or 1,225 gallons (4,630 liters) each year, just from one dripping faucet.

186. How concerned should I be about a leaky toilet?

A leaky toilet can waste as much as 200 gallons (260 liters) of water a day. A common reason toilets leak is the toilet flapper has become worn and no longer seals closed once the toilet has filled. Flappers are inexpensive rubber parts that can build up minerals or decay over time. If you are handy, it's easy to replace the flapper, and advice is available on the Internet. For more information, check out the USEPA's consumer website, "Fixing Leaks Around the Home," www.epa.gov.

187. I have a private well. Why should I conserve water?

Many aquifers in this country are seriously depleted, and groundwater sources are in danger. It can take anywhere from days to centuries for aquifers to recharge with new water. By conserving water, you can help maintain the source and avoid having your well drilled deeper to reach the aquifer.

188. Why do we still have a water shortage when it's been raining at my house?

It takes weeks to years for the water to recycle from rain to home use, depending on your water source. If your drinking water comes from a well, the rainwater needs to recharge into the ground before it affects the supply. If it comes from a reservoir or surface source, the supply may be far upstream from the rain at your house, so the water source isn't being replenished. Although the rain in your area may not help the water supply, it does lower the demand for water for watering lawns and gardens and washing cars.

189. During water shortages, shouldn't decorative fountains be turned off?

In most cases, fountain water is recirculated (used over and over) and is not wasted. If water losses from evaporation are high, however, fountains should be turned off.

190. Does not serving water to restaurant customers actually help us conserve?

Skipping water in restaurants serves as a good reminder to everyone about the importance of saving water, but the actual volume of water saved is small. The water that would be used to wash the water glasses is also saved, and this is usually more than the glass of drinking water served to customers (two glasses of water for each glass washed).

However, health experts emphasize the importance of drinking at least six to eight glasses of water each day, so if you want water to drink, ask for it. Skip it if you are just going to let the glass sit on the table while you drink something else.

191. Because the amount of water on the globe isn't changing, and the water in my area is plentiful, why should I conserve?

While you're right about the amount of water being constant, saving water saves you and your community money, so conservation is still important. For example, suppose you live in a growing community. As the population increases, so does the demand for water. This means that every so often, the water supplier must find another source of water, and in some areas, additional sources are hard to find. If everyone conserved, the water demand would not grow as fast, and the need to look for more water would be delayed. This permits the municipality to defer expenditures and to use the money for something else in the meantime. In addition, not all the water taken from the tap gets right back into the source. Aquifers, for example, take a long time to recharge and may never come back to their full capacity. Saving water can help your community, and it can also lower your water bill, as well as potentially lower your wastewater bill.

You can also look at it this way: When we use water, we generally add contaminants to it. Although the amount of water on the planet is constant, if we use the water for drinking, cleaning, and other domestic, commercial, and industrial purposes, these contaminants still must be removed.

192. What is graywater? Can we use it?

Graywater is used water from dishwashing, showers, sink use, or laundry water that has been used in the home—essentially any residential wastewater other than water from toilets. Graywater can be reused for other purposes, especially landscape irrigation because graywater can contain small bits of residue that can act as fertilizer. But always check with your health department because some states and localities regulate what you can do with the graywater.

The benefits of graywater recycling include

- lower freshwater use;
- less strain on failing septic tanks or treatment plants;
- less energy and chemical use;
- good source of irrigation water for plant growth; and
- reclamation of otherwise wasted nutrients.[1]

193. Our water utility has opened a reclaimed water plant in our town. What is reclaimed water? Can we drink it?

Reclaimed water, also referred to as reuse water or recycled water, is wastewater treated to reusable standards. Water users, whether industrial, commercial, or municipal, are increasingly being served with reclaimed wastewater in place of potable water at a cost substantially lower than the cost of developing new high-quality sources of supply for potable purposes. Reclaimed water is made suitable and safe for reuse through extensive wastewater treatment and by limiting public or worker exposure to the water through design and operational controls.[2]

Reclaimed water that is not treated to drinking water standards is sometimes distributed through a separate piping system. You may have seen the purple pipes. It is primarily used for irrigation of public parks, golf courses, cemeteries, plant nurseries, medians, and, if the system is reconfigured for hydrants, firefighting. Other uses may include industrial cooling and heating, toilet and urinal flushing, car washes, and other services.

---

1 Oasis Design. 2009. Gray Water Policy Center. https://oasisdesign.net/greywater/law
2 AWWA. 2019 (4th ed.). M24, *Planning for the Distribution of Reclaimed Water*. Denver: AWWA.

Indirect potable reuse is the practice of treating wastewater to drinking water standards and then adding it to a source of water such as a lake, river, or aquifer. A few utilities in the United States are practicing *direct potable reuse*—treating wastewater that is then returned to a drinking water system. California was the first state to approve this practice in 2023, and Texas and Arizona also approveof this practice. Other states have issued guidelines but implementation of this practice has yet to become widespread.

194. Is reclaimed water regulated?

The USEPA published guidelines for water reuse in 2012, but there are no federal regulations governing water reclamation and reuse in the United States. Several states, however, have established regulations or criteria regarding this water. Do be assured that any reclaimed water that is used for human consumption has to meet minimum safety standards and is cleaner than the water that is discharged directly from a wastewater plant into the environment.

If wastewater can be used productively, the savings in wastewater treatment can be passed on to users, making water reclamation and dual distribution systems more economically attractive.

Chapter 7

Regulations, Reporting, and Water Security

Providing safe drinking water and sanitation for all cannot be merely a slogan but rather a call to action so that future generations will have a drop to drink.
— Jack Hoffbuhr, former executive director of AWWA

195. Is the quality of drinking water regulated?

In the United States, the Safe Drinking Water Act (SDWA) protects the quality of drinking water supplied by public water systems (PWSs). It is administered by the US Environmental Protection Agency (USEPA). It was first passed in 1974 and expanded and strengthened several times since then. The SDWA establishes enforceable health-based limits on foreign matter in water, limiting everything from man-made chemicals and disease-causing microbes to water additives, such as fluoride and chlorine. Amendments to the SDWA are sometimes called "Rules," such as the Ground Water Rule (GWR) and the Lead and Copper Rule (LCR).

About 15% of the US population relies on private wells for drinking water. Private wells are often required by states to be tested for microbial contamination when they are first installed but are not monitored after that. These wells are not regulated by the USEPA and are not required to meet USEPA standards.

In Canada, drinking water is a shared responsibility of provincial, territorial, federal, and municipal governments. The federal government sets water quality guidelines and regulates federal facilities. The provinces and territories generally are responsible for regulating the provision of safe drinking water to the public, while municipalities usually deliver water to the public. The federal agency Health Canada uses its scientific and technical expertise to develop *Guidelines for Canadian Drinking Water Quality* in partnership with the provinces and territories. These guidelines are used by every jurisdiction in Canada and are the basis for establishing drinking water quality requirements for all Canadians.

196. What about state or provincial regulations?

States implement and enforce the requirements of the SDWA and, in some cases, implement more stringent requirements for public water supplies in their jurisdiction. Only the state of Wyoming and the District of Columbia have not accepted responsibility for enforcing the SDWA, so the USEPA has direct jurisdiction over the water systems in those localities.

Each state must adopt drinking water quality standards that are at least as strict as those of the federal government. The responsible state agency determines whether the PWSs in their state comply with all the requirements of the SDWA and takes action to ensure that those out of compliance improve their operations to ensure safe drinking water for their customers. In Canada, a provincial department, typically the department of health, environment, or municipal affairs, establishes regulations and guidelines pertinent to that province. The provinces have more control than the federal agency in regulating the water in Canada.

197. Is water that meets government drinking water standards absolutely safe?

Safety is relative, not absolute. For example, an aspirin or two may help a headache, but if you took a whole bottle at once, you'd probably become very ill. So, is aspirin safe? When setting drinking water standards, regulatory agencies use the concept of reasonable risk, not risk-free. Water with a complete absence of risk is not technologically feasible and would cost too much to produce. So, the answer to the question is no, drinking water isn't absolutely safe for all people, particularly those with compromised immune systems. But the likelihood of getting sick from drinking water that meets the federal standards is very small, typically one chance in a million.

One difficulty federal agencies have when trying to determine reasonable risk relates to the people who are more susceptible to getting sick than others, called the susceptible population. For example, only babies three months old or younger are seriously affected by nitrates in drinking water, so for that contaminant, they are the susceptible population: They are susceptible to getting sick from too much nitrate in their drinking water. The standard for nitrate, therefore, was chosen to protect these infants. With other contaminants, identifying the susceptible population is not as easy. Are they babies, elderly people undergoing cancer treatment in nursing homes, people who are positive for HIV (human immunodeficiency virus), or others? For each standard, the federal regulatory agencies must balance the risk to all these groups against the cost of treatment and arrive at a standard that will protect as many people as possible and still be affordable. This is called "the greatest good for the greatest number."

198. How do regulatory agencies set the standard for a chemical in drinking water?

In the United States, the National Institute of Environmental Health Sciences through the National Toxicology Program and USEPA laboratories conduct extensive tests on rodents with the chemical in question to determine its effects. Because rats and mice digest their food in the same way humans do, this information is extrapolated to determine a *reasonable risk* of exposure to the chemical over time. For most potentially cancer-causing chemicals, reasonable risk is defined as follows: If 1 million people drank water for a period of 70 years with the amount of chemical in it equal to the standard, no more than one additional person would be likely to get cancer from the drinking water—a very small risk. If possible, a safety factor is added that lowers the allowable exposure even further to determine the drinking water standard.

The federal agencies also conduct studies on appropriate treatment and monitoring technologies and work with water systems to collect contaminant occurrence data, which are used to conduct a cost–benefit analysis that both determines if the contaminant must be managed and determines what management actions are cost-effective. This provides the appropriate level of public health protection given the nature and distribution of the contaminant.

In Canada, the Federal–Provincial–Territorial Committee on Drinking Water (CDW) has been charged with identifying, assessing, and evaluating potential contaminants. The CDW uses risk assessments from both the USEPA and the World Health Organization for establishing standards for the national guidelines, along with regional research information on the occurrence of specific parameters.

199. How do we know what the allowable level of a contaminant is?

Once a drinking water standard has been chosen for any contaminant, it is called the maximum contaminant level, abbreviated MCL, for that contaminant. In addition to the MCL, by law, the USEPA must choose a maximum contaminant level goal (MCLG) for any regulated contaminant. This is the level below which there is no known or expected risk to health. The MCLG also contains a factor of safety. This is different than the reasonable risk concept, which is used to set MCLs. The USEPA does not enforce MCLGs. The USEPA's website lists the MCLs and MCLGs of all contaminants that are regulated in drinking water (www.epa.gov).

In Canada, the *Guidelines for Canadian Drinking Water Quality* is just that—guidelines—and unless a province adopts the guidelines as enforceable requirements, the allowable level is just a recommendation. Health Canada does set a maximum acceptable concentration (MAC) for parameters believed to have health effects. Check with your local utility if you have concerns with what the level of a given parameter in your water is.

200. Most federal standards are written like this: "Selenium—0.05 mg/L." What does "mg/L" mean?

The abbreviation mg/L stands for milligrams per liter which describes the concentration of a substance in water. In metric units, this is the weight of a chemical (selenium in the example) dissolved in 1 liter of water. In water, 1 mg/L is about the same as 1 part per million, or simply like the saying "one in a million." One liter is about equal to 1 quart, and 1 ounce is equal to about 28,500 milligrams, so 1 milligram is a very small amount.

For another example, a grain of salt weighs about 0.06 milligrams. If one grain of salt were added to one 1 liter of water, the concentration would be 0.06 mg of salt in the liter of water, written as **0.06** milligrams in each liter, or 0.06 mg/L. In the metric system, weights can be written in increasingly smaller amounts, each a thousand times smaller. For example, a 0.06-milligram grain of salt could be written as **60** micrograms (μ) of salt, and in the metric system of weights, a nanogram (n) is one thousand1,000 times smaller than a microgram (μ), and a picogram(p) is one thousand1,000 times smaller than a nanogram.

One milligram per liter (1 mg/L) is hard to imagine. Here are some examples: 11.6 days contain 1 million seconds, so 1 second out of 11.6 days is one in a million; or 25 grains of sugar in a quart of water is about 1 mg/L; or one drop in a 55-gallon drum is about 1 mg/L.

201. What is the Clean Water Act?

Officially known as the Federal Water Pollution Control Act, the Clean Water Act (CWA) was signed into law in 1972 by the US government to control the treatment of wastewater and its subsequent release into the environment. Under the CWA, any facility that intends to discharge waste materials into the nation's waters must obtain a permit before initiating a discharge. The intent of the CWA is to restore and maintain the chemical, physical, and biological integrity of the nation's waters by preventing pollution through the setting of standards for the treatment of discharges and providing assistance and guidance to wastewater treatment facilities and other industries that discharge to the waterways.

The CWA also requires states to develop management plans to protect waters from nonpoint sources, such as from agricultural runoff.

In Canada, the Fisheries Act protects water bodies from pollution, and each province and territory develops its own legislative framework for regulating water.

202. How do regulatory agencies choose which contaminants to regulate?

The USEPA has developed a Contaminant Candidate List (CCL) that identifies contaminants that are not regulated under the SDWA but are known or anticipated to occur at PWSs and may warrant regulations in the future. Certain water systems are required to monitor for certain unregulated contaminants under the Unregulated Contaminant Monitoring Rule (UCMR). The USEPA then analyzes contaminant occurrence, health effects, and other factors to determine whether a contaminant should be regulated. Canadian standards are based on similar occurrence and health effect studies.

203. If most tap water is safe, why are engineers and scientists still doing so much research, and why is the government thinking about more regulations?

Science is always improving, and monitoring tests and equipment have become more sensitive. This continued improvement leads to detection of contaminants in smaller and smaller amounts and leads to the discovery of unknown contaminants at minute concentrations. Medical research plays a part by determining whether these contaminants are a risk to human health. As new contaminants emerge that need to be researched, numerous organizations that focus on water research will conduct studies that will be shared with water utilities, health professionals, and regulators to determine what course of action should be taken to control the contaminant. Per- and polyfluoroalkyl substances (often called PFAS) and microplastics are examples of contaminants that are of concern because they have just recently been detected in drinking water, and the health implications of the chemicals are still unknown.

Drinking water suppliers are committed to keeping water safe for consumers and do so by complying with the SDWA. The CCL is used to prioritize research and data

133

collection efforts to help determine whether a specific contaminant needs to be regulated because of its health effects and/or occurrence levels.

204. Must all surface water supplies be filtered?

Since 1993, the USEPA has required that all US water suppliers using surface water include a filtration step during treatment. Waivers to this rule have been allowed if a water system can demonstrate exceptional water quality and show that source water quality can be protected by implementing strict watershed protection measures. Since the late 1990s, however, the USEPA has been requiring some systems, such as New York City, Boston, and Seattle, with filtration waivers to build filter plants. If you are interested in your local situation, talk to your water supplier.

In Canada, a similar situation exists. General policy requires filtration of surface water, but a local supplier that can demonstrate it meets stringent testing, reporting, and quality criteria need not filter. As in the United States, many localities are assessing the need for filtration.

205. What is the Ground Water Rule?

The US Ground Water Rule (GWR) was established under the authority of the SDWA in 2006 to protect the public from pathogen contamination in public drinking water systems that use groundwater for the source of water. Historically, groundwater has been considered to be pathogen-free, but some groundwater sources are contaminated and treatment is necessary. The GWR applies to all systems using groundwater, but it does not require all groundwater systems to treat their water. Under the rule, water utilities must identify any problems that could cause contamination and address those problems.

206. What is the Lead and Copper Rule?

The Lead and Copper Rule was authorized by the SDWA in 1991 and has been revised many times as the dangers of lead in drinking water have become better understood. In 2024, the Lead and Copper Rule Improvements (LCRI) were issued. This latest amendment of the 1991 Rule clarifies how and where monitoring of tap water is to be done, increases the requirements for customer education, and mandates the removal of lead bearing material, such as lead service lines.

Lead was regulated in the original SDWA of 1976, with an MCL set at 50 micrograms per liter ($\mu g/L$). But the Lead and Copper Rule of 1991 changed how lead in drinking water was regulated. Under the 1991 Rule, PWSs were required to sample at customer taps, and 90% of the results of that testing had to be less than an action level—15 $\mu g/L$ for lead and less than 1,300 $\mu g/L$ for copper. The LCRI of 2024 aims to reduce the action levels in the coming years. The number of customer taps sampled varied by the number of people served by the water supplier. The testing must be done every three years. The limits for lead and copper in drinking water are calculated differently than for any other contaminant. Only 90% of the samples must have levels less than the action level. For other regulated contaminants, all samples must have amounts less than the MCL.

207. How is my water tested, and who tests it?

All PWSs in the United States and Canada must test the treated water for nearly 100 parameters a specified number of times each year. Tests for microbes are conducted most often; the frequency varies depending on the population served by a water supplier. Rather than test for the actual pathogens, water systems test regularly for indicator microbes that are more readily collected and analyzed and that provide a good measure of whether the system is adequately protecting public health. The reasoning is that if

indicator organisms are present, germs are likely to be present also.

In the United States, these tests are conducted in state-certified laboratories using federally approved methods, some of which are quite complex. In Canada, the testing labs are nationally accredited and, in most cases, must be approved by the local or regional government, such as the province or First Nation. Private wells may be tested in connection with the sale of a home, but it is up to individual homeowners to test their wells on a regular basis. State health agencies can direct private well owners to appropriate laboratories for testing.

208. Can water systems be excused from monitoring for some contaminants?

In some limited cases, yes. These are called "waivers" or "exemptions" and they can be issued by the USEPA or by the state. Over time, by reviewing test results and keeping a watchful eye on potential problems, PWS s come to understand the likely threats to their water supplies. If a water utility does not have water quality problems, it can apply to its regulator for permission to test less frequently for certain contaminants. If state regulators agree that human or natural activities are unlikely to contribute that contaminant to the system's water, the request to avoid unnecessary testing may be granted. A waiver from some monitoring requirements in no way reduces the water supplier's responsibility to provide high-quality drinking water.

209. How do I find out if my water is safe to drink?

If your water comes from a PWS, contact your water supplier or local health department and ask if the water meets federal, state, or provincial standards. If it does, the water is safe to drink. This is a good practice when you move to a new location.

All public water utilities in the United States are required by federal law to provide customers with a water quality report called a consumer confidence report (CCR). Required by the 1996 amendments to the SDWA, and updated in 2024 as the Consumer Confidence Report Rule Revisions, the report must provide customers with basic information about their drinking water, including its source; susceptibility to contamination; level or range of levels of any contaminant found in the water, the MCL for that contaminant, and the likely source of that contaminant; the potential health effects of any contaminant detected that is greater than the MCL and the utility's actions to mitigate the contamination in tap water; and other information. Water systems serving more than 10,000 people are required to distribute reports twice per year, smaller systems once per year.

The CCR is usually included in a water bill as a paper copy or a link to an electronic copy and can usually be found on the water utility's website.

The USEPA also has compiled information about local water systems on its website (www.epa.gov) and provides information on how to find the CCR for any PWS.

Canadian law requires water quality problems to be reported to public health agencies but does not require violations to be reported to consumers.

If you have your own private water source, such as a well or spring, you are responsible for having it tested yourself. Once it has been tested, you can discuss the results with your local health department.

210. What information should I look for in the water quality report from my supplier?

Most reports contain a table of constituents found in the local drinking water. For each constituent, the table usually shows the USEPA's MCL, and the amount found in your drinking water. If the amount in your supply is the same as or less than the MCL, your supply is all right. This will be the case in most situations.

Note that the amounts of some constituents may be listed as "not detected" (ND) or "below detection limits" (BDL). This is not the same as zero, because a level of zero would mean that not any of that constituent was in the water—not even one molecule. Because testing instruments cannot measure that small an amount, only the smallest amount of material that will cause a reading is known. If no reading is obtained, any trace of that constituent in the sample was too small to register. The instrument operator then reports ND or BDL for that constituent.

If the MCL for some constituents is listed as "treatment technique" instead of a number, the utility is required to install and properly operate water treatment processes that reliably remove certain contaminants.

211. I've received a notice from my water utility telling me that something is wrong. What's that all about? What is a boil-water order?

In the United States, federal law states that water suppliers must promptly inform consumers if the water has become contaminated by something that could cause immediate illness. If such a violation occurs, the supplier has 24 hours to notify the public through the media or personal contact. The notification will also include information about whether the water should be boiled (called a "boil-water" order) to kill organisms before use, whether it is safe to use for nonpotable purposes such as bathing, and what actions are being taken by the water supplier to solve the problem. This is also common practice in Canada.

If you or someone in your household is immunocompromised, elderly, or very young, you may want to take extra precautions when there is a water alert because these people may be more vulnerable to contaminants in drinking water than the general population.

All violations are important, of course, but they are not all equally important. For instance, if the problem reported to customers is that the water supplier has not sampled the water as frequently as required by the regulations, this does not necessarily mean that the quality of the water is poor. One violation involving insufficient sampling, improper reporting, or substandard water quality does not mean the water is unsafe to drink. It does mean, however, that the water utility should improve its operations.

Remember, if you have a boil-water order in your area, be sure to throw all of your ice cubes or any beverage made with water during the contamination period away. They may have been made with contaminated water.

212. I live in an apartment and don't get a water bill. How will I know if there are any problems with my tap water?

Serious problems, called acute violations, must be broadcast on the radio and TV and published in general circulation newspapers. Less serious problems must be published in general circulation newspapers and may also be otherwise posted or be sent to everyone who pays a water bill—in your case, the apartment owner or manager.

Ask your apartment manager to post anything included with the water bill. Even when there are no problems, water suppliers often include valuable information (often called bill stuffers) with the water bill. Apartment dwellers should be able to see this information. You can also read the CCR (discussed previously in this chapter) on the water utility's website. If you don't know which PWS supplies your apartment complex, start by calling the local city public works department and they can tell you where your water comes from.

213. What are the chances of a water system being the object of a terrorist attack?

Just like many public facilities, water systems are considered by the federal government as a possible target of terrorist activity. Because of this, the Bioterrorism Preparedness and Response Act of 2002 (Bioterrorism Act) required all community water supply systems in the United States to conduct a risk assessment to determine what facilities and operations were most vulnerable to an attack. After the assessment, many utilities took actions to protect what they identified as their most vulnerable assets, be it the treatment facility, water distribution system, computer programs, or other components of the water system. Also, America's Water Infrastructure Act of 2018 requires community water systems serving more than 3,300 people to develop or update risk assessments and emergency response plans (ERPs).

There are several federal initiatives through the Department of Homeland Security that have provided guidance to utilities in implementing security measures. Many water utilities are also continually monitoring threat activity and are part of a confidential national security alert program called the Water Information Sharing and Analysis Center (WaterISAC). To coordinate protection from and awareness of threats to water and wastewater infrastructure, the Water and Wastewater Government Coordinating Council and Sector Coordinating Council (WGCC) provides a forum for water utilities to communicate. Information on that initiative can be found on www.cisa.gov.

Risks to water utilities can include physical threats, contamination threats, and cybersecurity threats.

214. What would happen if somebody intentionally put a dangerous chemical in our water?

Because water utilities are aware of the potential for attacks on water supplies, most have increased their monitoring and testing of water quality to provide early warning of possible threats. Many water quality parameters can be measured using on-line instrumentation, which continuously measures and records these parameters and alerts operators if unusual measurements are detected. Conductivity, pH, and turbidity are common parameters that could change noticeably if something strange was added to the water. The operators would then conduct more testing to determine what the added substance was and take precautions to prevent the contamination from becoming a problem at the tap. Customer complaints of unusual water appearance, tastes, or odors to the water utility can also help identify problems.

If the contamination had already reached the distribution system, the water system would alert the end users through public notification systems and possibly issue do-not-use or boil-water orders. The utility would then have to decontaminate the system, which would require shutting down the distribution system while it is cleaned and disinfected. In that case, the utility would provide consumers with bottled water or distribute clean water from trucks or large containers called water buffaloes.

215. What about the danger of a chemical from a water utility being released into the air?

Just as with other security measures, utilities that use potentially dangerous chemicals, such as chlorine, have conducted risk assessments to determine what the effects of a chemical release would be and have taken precautions to safeguard their hazardous materials. This includes shoring up containment facilities, detecting any dangerous chemical release early, "scrubbing" equipment that would

141

remove the chemical from the air before it can be released into the community, and ensuring the safe delivery of hazardous materials.

Also, more and more utilities are switching to alternative disinfectants that are less hazardous.

216. Are water utilities prepared for natural disasters such as hurricanes and floods?

The risk and resilience assessment (RRA) addresses every aspect of the water system from source to tap, including physical, electronic, financial, and operational components. The ERP must include plans, procedures, actions, and equipment to address each item identified in the RRA. Both the RRA and the ERP must be reviewed and updated every five years.

That said, there is always the chance that something will happen that badly damages a facility and prohibits utility personnel from working. If this were to happen, a utility could receive assistance from other water suppliers in neighboring communities, or even other states, through a program called WARN, which stands for Water and Wastewater Agency Response Network. AWWA (www.awwa.org) coordinates the national WARN program and provides resources for water sector employees to help utilities affected by natural disasters. In the aftermath of Hurricanes Helene and Milton in 2024, for example, water system operators from across the United States assisted water systems in Florida and North Carolina.

Chapter 8

Fascinating Facts

Water is unique in that it is the only natural substance that is found in all three states—liquid, solid, and gas at the temperatures normally found on Earth. Earth's water is constantly interacting, changing, and in movement.
—US Geological Survey, *Water Science for Schools*, 2008

LIQUID

217. How much does water weigh?

One US gallon (3.8 liters) of water weighs about 8.3 pounds (3.8 kilograms).

218. I have a typical residential lot of a quarter acre. If it rains 1 inch (25 millimeters), how much water falls on my lot?

About 7,000 US gallons (26,500 liters), or nearly 30 tons (27 metric tons) of water. That is equal to 0.02 acre-feet.

219. Why does drinking water get bubbles in it after it has been left out for a few hours?

Bubbles form in drinking water when gases, which have been dissolved in the water, are released into the water. The gases are usually nitrogen and oxygen, which can come from the air or from water treatment processes. When there is more air in the water than the water can hold, the dissolved air comes out of the water and forms bubbles. The amount of gas that water can hold depends on temperature and pressure. If the water in the glass is warming up, gas is released because warm water cannot hold as much gas as cold water. Also, if the water in the glass is under less pressure than it was before it came out of the tap, gas will be released.

220. Why does water swirl when it goes down the drain?

It's a common misconception that the earth spinning on its axis—the *Coriolis effect*—causes draining water to spin one direction in the Northern Hemisphere and another in the Southern Hemisphere. The rotation of the earth is a major factor in the motion of ocean currents and weather, but your bathtub or sink is too small to be substantially affected by this motion.

Instead, the swirling at the drain is caused by the same effect as the motions of a spinning ice skater. When the skater pulls their arms in close to the body, the skater spins faster. Water in a basin is never quite quiet, so when the water is pulled toward a drain, it's drawn down at an angle, and any residual rotation becomes larger on the broad surface and increasingly smaller toward the narrower drain opening, gaining momentum so you can see it, like the ice skater pulling their arms close to their body.

221. Why is water in the ocean blue?

Actually, water has no color! Sunlight contains all colors of the rainbow, but when you see a color, it is because that particular color of light bounces off the object and reaches your eye, and the other colors are absorbed by the pigment of the object. The sky is blue on a nice day because only blue light reflects off the particles in the atmosphere. The ocean is blue because as the blue light comes out of the sky, it reflects from the water and into your eye, fooling your eye into thinking the water is blue. The ocean absorbs red light, which leaves more blue light to reflect off the sea. Next time the sky is overcast, notice how the ocean loses most of its color because so little blue light is coming out of the sky.

Where the ocean is clear and shallow, the sea bottom can also influence the color, so the water will reflect different shades of blue, and often green. And the ocean can also appear to be other colors if runoff or turbulence has suspended material in the water. Swimming pools are usually painted blue to make the pool water appear blue.

222. Why is ocean water salty?

Ocean water contains a significant amount of dissolved sodium and chloride—salt—that comes from rivers and streams carrying tiny bits of rock that the flowing water has eroded on its way to the ocean. The erosion of the ocean floor also contributes to the salt content. Seawater is about 3.5% salt.

223. Why, then, aren't big lakes that are fed by rivers and streams salty?

All water has some salt content, although most taste buds are not sensitive enough to detect it. Freshwater lakes are replenished at a relatively frequent rate, and rivers are fed by fresh snowmelt, rain, and underground springs, which keep salt from accumulating in the water.

224. Should I wash my hands with soap in hot water?

The Centers for Disease Control and Prevention (CDC) says that the temperature of the water does not appear to affect germ removal. The CDC suggests wetting your hands with clean, running water (warm or cold), turning off the tap, and applying soap. Using soap to wash hands is more effective than using water alone because the surfactants in soap lift soil and microbes from skin and scrubbing hands more thoroughly when using soap further removes germs.

225. Is it true that a microwave oven heats food and beverages by flipping water molecules back and forth? How is that possible?

A water molecule has one oxygen atom with two hydrogen atoms connected to it—thus, H_2O. The shape of the molecule is like a flat V, with the oxygen at the point and the two hydrogen atoms hanging off it. Oxygen has two negative charges, and each hydrogen atom has a positive charge, so think of a water molecule as a little magnet, with one end negative and the other end positive. When the microwave is on, the waves affect the way the water molecules line up. Because the microwaves are produced by alternating current (AC), they change direction about 2 billion times each second. Each time the microwaves change direction, water molecules are flipped, too. Because water molecules are so small, trillions upon trillions of molecules are in any food or beverage you put in the microwave. All these molecules rapidly flipping back and forth create heat for defrosting, cooking, and warming food and beverages.

226. I've heard of water exploding after it comes out of the microwave, causing serious burns. What can I do to avoid this?

Microwaved water and other liquids do not always bubble when they reach the boiling point. This happens when the water container has perfectly smooth surfaces and the water heats faster than the vapor bubbles can form. This superheated liquid will bubble up out of a cup when it is moved or when something is put into it, such as a spoon or tea bag. It is a rare occurrence, because most containers contain impurities that disrupt the perfect molecule alignment that is required for this problem to occur.

To prevent this from happening, do not heat any liquid for more than 2 minutes per cup. After heating, let the cup stand in the microwave for 30 seconds before moving it or adding anything to it. Or you can put a wooden stir stick (no metal) in the water before you heat it.

227. Why does it take longer to cook things in boiling water when I am camping in the mountains?

The temperature at which water boils depends on the atmospheric pressure. Atmospheric pressure depends on the weight of the air above any location. At sea level, the weight of the atmosphere "pushes down" with a pressure of 14.7 pounds on 1 square inch (101.3 kilopascals), and water boils at 212°F (100°C). Go up into the mountains, however, and the atmosphere is thinner above you, with less pressure. The lower the pressure, the lower the temperature at which water boils. The temperature at which water boils goes down about 2°F (1°C) for each 1,000 feet (305 meters) of elevation. If you are camping in the mountains at 10,000 feet (3,050 meters), water boils at only 201°F (94°C). At this lower temperature, food takes longer to cook.

228. Are raindrops really shaped like teardrops?

No, raindrops aren't shaped like teardrops. Small rain-drops (with a diameter of less than 0.1 inches [0.25 centi-meters]) are shaped like a sphere; larger ones are shaped more like a hamburger bun. When raindrops become larger than a diameter of 0.4 inches (1 centimeter), they distort into a shape like a parachute with a tube of water around the base, and then they break up into smaller drops.

229. Why is water referred to as "the universal solvent"?

Water is called the universal solvent because it can dis-solve more substances than any other liquid. This is impor-tant to every living thing on earth. It means that wherever water goes, through the air, the ground, or our bodies, it takes along valuable chemicals, minerals, and nutrients. Think of the Grand Canyon and how the flow of the Colora-do River helped create it. A water molecule's simple, stable bond of two hydrogen atoms and one oxygen atom creates a neutral solution that breaks the bonds of larger, more com-plex molecules.

230. I saw a TV demonstration in which a container was slowly filled with water, and the water rose to a level above the top of the container without spilling. How is this possible?

Water molecules at the surface form a thin skin on the top of the water, and this skin is what holds the water above the top of the container. This is called *surface tension*. If you break the skin, the water will spill down the side of the container.

Try this yourself. Fill a container partially with water. When the surface is still, carefully lay a sewing needle on the surface. The skin will hold the needle up and allow it to float on the surface.

231. Why can't I pick up a piece of dirt from the floor with a dry finger, but if I wet my finger the dirt will stick to it?

This is a result of the same property of water as noted in the previous answer. The surface tension of the liquid on your finger causes the dirt to cling to your finger.

SOLID

232. Why does ice float instead of sink?

Frozen water is less dense than liquid water because as water freezes, the hydrogen bonds connecting different water molecules line up to keep the negatively charged oxygen atoms apart, creating equal spaces between the molecules and freezing the water in a lattice-type formation. This extra air between the molecules creates buoyancy that allows ice to float.

If this were not the case, ice would sink when it was formed and go to the bottom of all the lakes, ponds, and reservoirs. Even icebergs would sink to the bottom of the ocean. When spring came, not enough heat would get to the bottom to melt the ice down there. So, after a while, almost all the water on the earth would be frozen, and life as we know it would not have developed on the planet.

233. How do ice skates work to allow the skater to glide over the ice?

A thin film of liquid water covers the surface of any body of ice. How thick the liquid layer is depends on the surface temperature—the colder the ice, the thinner the layer. An ice skate blade bears the weight of the skater on a very small area, and when the skate blade passes over the ice, the increase in pressure and kinetic friction raises the temperature and melts the ice under the blade, creating an even thicker layer of water that has little resistance to the ice skate blade.

234. I saw a science experiment where one block of ice was hung from a wire looped around it, and another ice block was hung from a plastic line. Each had the same amount of weight hanging from it. The weight pulled the wire through the block of ice but not the plastic line. What was going on here?

Wire is a better heat conductor than plastic, and it absorbs the heat from the ice that melts and refreezes under pressure from the weight. Plastic doesn't absorb and conduct heat nearly as well, so considerably less melting occurs.

235. What makes the ice cubes I make at home cloudy?

Commercially made ice is stirred as it is being frozen; household ice is not. Without mixing, many more ice crystals form, and air is trapped in the ice. Light rays are distorted by these crystals and air, and this distortion gives home-frozen ice a cloudy appearance.

236. Why do ice cubes bulge from the top of the ice cube tray?

Unlike most things, which get smaller when they get colder, water gets bigger (expands) by 9% when it freezes. Because an ice cube tray has a bottom and four sides that don't move, ice bulges out of the open top when the water expands.

Because frozen water (ice) has expanded, it is lighter than water. Therefore, in the winter, ice floats on the surface of a body of water while the water underneath stays liquid, providing organisms, including fish, with a place to survive during the cold weather.

237. Snow, sleet, freezing rain, and hail—how are they different?

All precipitation begins as ice crystals in the upper atmosphere.

Snow falls through cold atmospheric layers that are consistently below freezing, giving the snowflake no opportunity to melt.

Sleet melts on the way down and refreezes into icy pellets before hitting the ground. Sleet usually occurs during the winter and is smaller than 0.30 inches (0.76 centimeters) in diameter.

Freezing rain, or drizzle, is snow that melts and stays liquid on the way down (like regular rain) but freezes after hitting the ground or other frozen surfaces. Freezing rain is associated with ice storms that coat trees, power lines, and roads.

Hail is different. It is created during strong thunderstorms that contain strong updrafts. As the precipitation from a circular updraft first nears the ground, it is blown back up into the cold layer of sky, where it picks up more moisture that freezes on the outside, making the hailstone a bit bigger. Then it starts down again. In very strong storms, the hail goes back up several times, getting bigger each time. When it finally gets so heavy that the updrafts cannot blow it upward anymore, it falls to the ground. If you pick up a large hailstone and cut through it, you will see rings as on a cut tree trunk, each ring indicating another trip up into the sky. Soft hail is called "graupel" or snow pellets. Hail is usually round, is at least 0.20 inches (0.51 centimeters) in diameter, and occurs primarily in the spring and summer.

238. Is it true that all snowflakes are different, no two being exactly alike?

The eventual shape of the snowflake depends on the temperature of the original ice crystal, the temperature zones it falls through, and how it falls, which is when it picks up additional crystals to become a flake. A perfectly symmetrical snowflake spins like a top on the way down.

Each snow crystal contains about 1 quintillion (1,000,000,000,000,000,000) molecules of water. Each molecule is slightly different, and all molecules are scattered randomly throughout the crystal, creating even more ways for the crystal to be different. Then, consider that the crystals fall at different rates and movements—spinning, arcing, flipping—depending on the wind currents and the various temperature changes they encounter as they fall. Some even pick up bits of dirt and debris in the air. All of this means it is unlikely that anyone would see two identical snowflakes.

239. Even when the temperature stays below freezing, the depth of the snow on the ground decreases. How is this possible?

When snow falls from the sky, it settles in layers that are full of air around the snowflakes. As the snowpack settles, it pushes some of the air out, and the snow on the ground becomes more compact. Also, wind and sun can affect the amount of snow on the surface, in a process similar to evaporation (liquid turning to gas), but known as sublimation (frozen matter turning to gas). Even if temperatures aren't warm enough to melt the snow, some will dissipate into water vapor, further decreasing the snowpack.

240. I saw a demonstration where someone put a black cloth on a snowbank on a cold, sunny day and the cloth sank down into the snow. Why did that happen?

The sun sends heat and light to the earth as radiation. Many things reflect light into our eye, so we see them. Something black does not reflect any light, so we see it as black (the absence of reflected light). The same thing is true of heat. The black cloth does not reflect radiation and thus keeps both the light and the heat. This causes the black cloth to get warm, so it melts the ice beneath it. A white cloth would not melt the snow because it reflects both the light and heat, staying cool. Some people use a black hose to warm swimming pool water in sunny areas. The black hose lies out in the sun, collects the heat from sunlight, and warms the water being pumped slowly through it.

241. Why is the ice in glaciers blue?

The weight of the glacier produced tremendous pressure on the ice for a long, long time, which changed the nature of the tiny ice crystals so that they reflect only short, blue-light wavelengths, while absorbing long, red wavelengths. The longer the path that light must travel through the ice, the bluer the ice appears. This does not happen when you make ice in the refrigerator because this ice is not under high pressure.

242. We often use artificial cold packs in our coolers to keep things cold. How do they work?

Commercial ice packs are about 98% water, but they have added chemicals—usually ammonium chloride—to lower the freezing point. Because the chemical reaction between the water and the chemical lowers the temperature of the pack to less than 32°F (0°C), the packs can absorb heat from other objects and thus keep food and drinks cold. This idea is not new. Before freezers were common, ice cream was made by surrounding cream-filled containers with water, ice, and salt. The added salt lowered the temperature of the surrounding mixture and drew heat out of the cream until it froze. When salt (sodium chloride) or calcium chloride is used to deice pavements or sidewalks, the same principle applies.

Reusable gel packs contain water mixed with a refrigerant gel that prevents the water from completely freezing, so the pack can conform to the shape of the object it is next to.

GAS

243. What is dry ice?

Dry ice is a solid form of carbon dioxide (CO_2). Unlike water, CO_2 does not have a liquid form and can only be a gas or a solid. Water freezes at 32°F (0°C), but because it is originally a gas, CO_2 must be much colder to form a solid. For CO_2 to become dry ice, it must be cooled down to −109.3°F (−78.5°C).

It is called dry ice because no liquid is involved. When dry ice warms up, it doesn't melt into a liquid but escapes into the air as a gas by the process of sublimation. Dry ice is often used to make smoke or fog in stage productions by placing it in a bucket of water, which creates low-sinking dense clouds of fog.

244. If we can't see air or the water vapor in it, why can we see steam from a boiling teakettle or a hot cup of soup?

It's true that pure vaporized water is a completely invisible gas. But the steam you see above a hot, wet surface is really a mist or vapor that occurs when the gas mixes with the cooler air, which slows the gas molecules down and combines them with other water molecules, condensing them and forming tiny visible droplets of liquid water.

245. Why can we see our breath when it is cold outside?

The effect is almost the same as a boiling teakettle; the air from your lungs is full of moisture, and when your warm breath hits the freezing air, it condenses into a visible liquid or, if it's very cold, into frozen crystals.

Appendix A:
Acronym List

ANSI	American National Standards Institute
AWWA	American Water Works Association
BDL	below detection limits
CCL	Contaminant Candidate List
CCR	Consumer Confidence Report
CDC	Centers for Disease Control and Prevention
CDW	Federal–Provincial–Territorial Committee on Drinking Water (Canada)
CO_2	carbon dioxide
CWA	Clean Water Act
CWS	community water system
DBP	disinfection byproduct
DEA	US Drug Enforcement Administration
EDC	endocrine disrupting compound
ERP	compounds emergency response plan
ETV	Environmental Technology Verification
FDA	U.S. Food and Drug Administration
HAA5	five haloacetic acids
HDPE	high-density polyethylene
IRIS	Integrated Risk Information System
ISAC	Information Sharing and Analysis Center
GAC	granular activated carbon
GWR	Ground Water Rule
kPa	kilopascals

$KMnO_4$	potassium permanganate
LCR	Lead and Copper Rule
LCRI	Lead and Copper Rule Improvements
MAC	maximum acceptable concentration (Canada)
MCL	maximum contaminant level (United States)
MCLG	maximum contaminant level goal
MIB	methylisoborneol
NTNCWS	nontransient noncommunity water system
ND	not detected
NRC	US Nuclear Regulatory Commission
NSF	National Science Foundation
PCE	perchloroethylene or tetrachloroethylene
PET	polyethylene
PFAS	per- and polyfluoroalkyl substances
PFOA	perfluorooctanoic acid
PFOS	perfluorooctane sulfonic acid
PFHxS, PFNA and HFPO-DA	generally known as GenX compounds
POE	point-of-entry
POU	point-of-use
PP	polypropylene
PPCPs	pharmaceuticals and personal care products
ppm	parts per million
psi	pounds per square inch
PVC	polyvinyl chloride
RO	reverse osmosis
RRA	risk and resilience assessment
SMCL	secondary maximum contaminant levels

SDWA	Safe Drinking Water Act
SMCL	secondary maximum contaminant level
TCE	trichloroethylene
TNCW	transient noncommunity water system
THMs	trihalomethanes
T&O	taste and odor
UCMR	Unregulated Contaminant Monitoring Rule
UV	ultraviolet
USEPA	US Environmental Protection Agency
USP	United States Pharmacopeia
WARN	Water & Wastewater Agency Response Network
WHO	World Health Organization
WRF	The Water Research Foundation
WQA	Water Quality Association

Appendix B:
Inorganic Constituents of Interest in Water

This material is adapted from: Sandeen, W. G., "Groundwater Resources of Rusk County, Texas," US Geological Survey, Open File Report 83-757 (1984). Note: Secondary regulations of the US Environmental Protection Agency mentioned below are nonenforced recommendations.

ALKALINITY

Description and Sources
Alkalinity is a measure of the capacity of a water to neutralize a strong acid, usually to pH of 4.2. Alkalinity in natural waters usually is caused by the presence of bicarbonate and carbonate ions and to a lesser extent by hydroxide and minor acid radicals such as borates, phosphates and silicates. Carbonates and bicarbonates are common to most natural waters because of the abundance of carbon dioxide and carbonate minerals in nature. The alkalinity of natural waters varies widely but rarely exceeds 400 to 500 mg/L as $CaCO_3$.

Effect on Drinking Water
Alkaline waters may have a distinctive unpleasant taste.

CALCIUM (Ca)

Description and Sources
Calcium is widely distributed in the common minerals of rocks and soils and is the principal cation in many natural

fresh waters, especially those that contact deposits or soils originating from limestone, dolomite, gypsum and gypsiferous shale. Calcium concentrations in freshwaters usually range from zero to several hundred mg/L. Larger concentrations are not uncommon in waters in arid regions, especially in areas where some of the more soluble rock types are present.

Effect on Drinking Water

Calcium contributes to the total hardness of water. Small concentrations of calcium carbonate combat corrosion of metallic pipes by forming protective coatings. Calcium in domestic water supplies is objectionable because it tends to cause incrustations on cooking utensils and water heaters and increases soap or detergent consumption in waters used for washing, bathing and laundering.

CHLORIDE (Cl$^-$)

Description and Sources

Chloride is relatively scarce in the earth's crust but is the predominant anion in sea water, most petroleum-associated brines, and in many natural freshwaters, particularly those associated with marine shales and evaporites. Chloride salts are very soluble and once in solution tend to stay in solution. Chloride concentrations in natural waters vary from less than 1 mg/L in stream runoff from humid areas to more than 100,000 mg/L in groundwaters and surface waters in arid areas. The discharge of human, animal or industrial wastes and irrigation return flows may add significant quantities of chloride to surface and groundwaters.

Effect on Drinking Water

Chloride may impart a salty taste to drinking water and may accelerate the corrosion of metals used in water-supply systems. According to the National Secondary Drinking Water Regulations of the US Environmental Protection Agency, the maximum contaminant level of chloride for public water systems is 250 mg/L.

DISSOLVED SOLIDS

Description and Sources

Theoretically, dissolved solids are dry residues of the dissolved substances in water. In reality, the term "dissolved solids" is defined by the method used in the determination. In most waters, the dissolved solids consist predominantly of silica, calcium, magnesium, sodium, potassium, carbonate, bicarbonate, chloride and sulfate, with minor or trace amounts of other inorganic and organic constituents. In regions of high rainfall and relatively insoluble rocks, waters may contain dissolved-solids concentrations of less than 25 mg/L; but saturated sodium chloride brines in other areas may contain more than 300,000 mg/L.

Effect on Drinking Water

Dissolved-solids values are used widely in evaluating water quality and in comparing waters. The following classifications based on the concentrations of dissolved-solids commonly is used by the U.S. Geological Survey.

Classification	Dissolved-solids concentration (mg/L)
Fresh	< 1,000
Slightly saline	1,000–3,000
Moderately saline	3,000–10,000
Very saline	10,000–35,000
Brine	> 35,000

The National Secondary Drinking Regulations (U.S. Environmental Protection Agency) set a dissolved-solids concentration of 500 mg/L as the maximum contaminant level for public water systems. This level was set primarily on the basis of taste thresholds and potential physiological effects, particularly the laxative effect on unacclimated users. Although drinking waters containing more than 500 mg/L are undesirable, such waters are used in many areas where less mineralized supplies are not available without any obvious ill effects.

FLUORIDE (F⁻)

Description and Sources
Fluoride is a minor constituent of Earth's crust. The calcium fluoride mineral fluorite is a widespread constituent of resistate sediments and igneous rocks, but its solubility in water is negligible. Fluoride commonly is associated with volcanic gases, and volcanic emanations may be important sources of fluoride in some areas. The fluoride concentration in fresh surface waters usually is less than 1 mg/L, but larger concentrations are not uncommon in saline water from oil wells, groundwater from a wide variety of geologic terrain, and water from areas affected by volcanism.

Effect on Drinking Water
Fluoride in drinking water decreases the incidence of tooth decay when the water is consumed during the period of enamel calcification. Excessive quantities in drinking water consumed by children during the period of enamel calcification may cause a characteristic discoloration (mottling) of the teeth. Thus, the EPA has an upper allowable limit for fluoride in drinking water.

HARDNESS

Description and Sources
Hardness of water is attributable to all polyvalent metals but principally to calcium and magnesium ions. Water hardness results naturally from the solution of calcium and magnesium, both of which are widely distributed in common minerals of rocks and soils. Hardness of waters in contact with limestone commonly exceeds 200 mg/L. In

waters from gypsiferous formations, a hardness of 1,000 mg/L is not uncommon.

Effect on Drinking Water

Excessive hardness of water for domestic use is objectionable because it causes incrustations on cooking utensils and water heaters and increased soap or detergent consumption.

IRON (Fe)

Description and Sources

Iron is an abundant and widespread constituent of many rocks and soils. Iron concentrations in natural waters are dependent upon several chemical processes including oxidation and reduction; precipitation and solution of hydroxides, carbonates and sulfides; complex formation, especially with organic material; and the metabolism of plants and animals. Dissolved-iron concentrations in oxygenated surface waters seldom are as much as 1 mg/L. Some groundwaters, unoxygenated surface waters such as deep waters of stratified lakes and reservoirs, and acidic waters resulting from discharge of industrial wastes or drainage from mines may contain considerably more iron.

Corrosion of iron casings, pumps, and pipes may add iron to water pumped from wells.

Effect on Drinking Water

Iron is an objectionable constituent in water supplies for domestic use because it may adversely affect the taste of water and beverages and stain laundered clothes and plumbing fixtures. According to the USEPA National Secondary Drinking Water Regulations, the maximum contamination level of iron for public water systems is 0.3 mg/L.

MAGNESIUM (Mg)

Description and Sources

Magnesium ranks eighth among the elements in order of abundance in Earth's crust and is a common constituent in natural water.

Ferromagnesian minerals in igneous rock and magnesium carbonate in carbonate rocks are two of the more important sources of magnesium in natural waters. Magnesium concentrations in freshwaters usually range from zero to several hundred mg/L; but larger concentrations are not uncommon in waters associated with limestone or dolomite.

Effect on Drinking Water

Magnesium contributes to the total hardness of water. Large concentrations of magnesium are objectionable in domestic water supplies because they can exert a cathartic and diuretic action upon unacclimated users and increase soap or detergent consumption in waters used for washing, bathing and laundering.

MANGANESE (Mn)

Description and Sources

In chemical behavior and occurrence in natural water, manganese resembles iron. Manganese is much less abundant in rocks, however. As a result, the concentration of manganese in water is generally less than that of iron. Under reducing conditions (the lack of oxygen dissolved in water) in water containing dissolved carbon dioxide, manganese dissolves as manganous ion. This sometimes occurs in groundwaters and in water near the bottom of lakes and reservoirs.

Manganous ion is more stable in water in the presence of dissolved oxygen than ferrous (reduced) iron under similar circumstances.

Manganese concentrations greater than 1 mg/L may result where manganese-bearing minerals are attacked by water under reducing conditions or where some types of bacteria are active.

Effect on Drinking Water

Manganese is an essential trace element for humans. It plays an important role in many enzyme systems. The main problem with manganese in drinking water has to do with undesirable taste and discoloration (black) of the water. The EPA secondary drinking water standard for manganese in drinking water is 0.05 mg/L to prevent these problems. Studies have shown that high levels of manganese can be harmful, but it is not clear how much an impact manganese in water has on human health. Canada has set a health based limit of 0.12 mg/L in drinking water.

NITROGEN (N)

Description and Sources

A considerable part of the total nitrogen of Earth is present as nitrogen gas in the atmosphere. Small amounts of nitrogen are present in rocks, but the element is concentrated to a greater extent in soils or biological material. Nitrogen is a cyclic element and may occur in water in several forms. The forms of greatest interest in water, in order of increasing oxidation state, include organic nitrogen, ammonia nitrogen (NH_3-N), nitrite nitrogen (NO_2-N), and nitrate nitrogen (NO_3-N). These forms of nitrogen in water may be derived naturally from the leaching of rocks, soils and decaying vegetation; from rainfall; or from biochemical conversion of one form to another. Other important sources of nitrogen in water include effluent from wastewater treatment plants, septic tanks and cesspools and drainage from barnyards, feedlots and fertilized fields. Nitrate is the most stable form of nitrogen in an oxidizing environment and is usually the dominant form of nitrogen in natural waters. Significant quantities of reduced nitrogen often are present in some groundwaters, deep unoxygenated waters of stratified lakes and reservoirs.

Effect on Drinking Water

Nitrate and nitrite are objectionable in drinking water because of the potential risk to bottle-fed infants for methemoglobinemia, a sometimes fatal illness related to the impairment of the oxygen-carrying ability of the blood.

pH

Description and Sources

The pH of a solution is a measure of its hydrogen ion activity. By definition, the pH of pure water at a temperature of 25°C is 7.00. Natural waters contain dissolved gases and minerals, and the pH may deviate significantly from that of pure water. Rainwater not affected significantly by atmospheric pollution generally has a pH of 5.6 because of the solution of carbon dioxide from the atmosphere. The pH range of most natural surface waters and groundwaters is about 6.0 to 8.5. Many natural waters are slightly basic (pH >7.0) because of the prevalence of carbonates and bicarbonates, which tend to increase the pH.

Effect on Drinking Water

The pH of a domestic water supply is significant because it may affect taste, corrosion potential and water-treatment processes. Acidic waters may have a sour taste and cause corrosion of metals and concrete. The USEPA National Secondary Drinking Water Regulations set a pH range of 6.5 to 8.5 as the maximum contaminant level for public water systems.

POTASSIUM (K)

Description and Sources

Although potassium is only slightly less common than sodium in igneous rocks and is more abundant in sedimentary rocks, the concentration of potassium in most natural waters is much smaller than the concentration of sodium. Potassium is liberated from silicate minerals with greater difficulty than sodium and is more easily adsorbed by clay minerals and reincorporated into solid weathering products. Concentrations of potassium of more than 20 mg/L are unusual in natural fresh waters, but much larger concentrations are not uncommon in brines or in water from hot springs.

Effect on Drinking Water

Large concentrations of potassium in drinking water may act as a cathartic, but the range of potassium concentrations in most domestic supplies seldom causes these problems.

SALINITY

Description and Sources

Salinity is the measure of the concentration of all dissolved salts in water. Salinity is a subset of dissolved solids, which measures all dissolved compounds in a water sample. Presence and concentration of salts in water depend on the surrounding environment. The concentration of dissolved salts in seawater (salinity) consists mostly of sodium and chloride—usually 85% of the total—but also can include magnesium, sulfate, calcium, potassium, bicarbonate, and bromine. Freshwater salinity has smaller amounts of some of the same salts as seawater.

Salinity can be expressed as a percentage (parts per hundred), such as 3.5% salinity for a seawater sample. Salinity in freshwater can also be expressed as a percentage, but it is more typically expressed as parts per million, or mg/L. In the seawater example, 3.5% salinity is equivalent to 35,000 mg/L. Good quality drinking water has less than 600 mg/L salinity (<0.1% salinity).

Most salinity comes from the erosion and dissolution of soil and rocks that come in contact with the water body. Some salinity is also introduced through activities such as agricultural runoff, industrial and mining processes, water softening, and road deicing.

Effect on Drinking Water
Too much salinity gives drinking water an unpleasant taste and potentially can be harmful for individuals with certain health conditions. Changes in salinity in drinking water sources can be harmful to the aquatic environment and result in more difficult treatment.

SILICA (SiO$_2$)

Description and Sources
Silica ranks second only to oxygen in abundance in Earth's crust. Contact of natural waters with silica-bearing rocks and soils usually results in a concentration range of about 1 to 30 mg/L; but concentrations as large as 100 mg/L are common in waters in some areas.

Effect on Drinking Water
Silica in some domestic water supplies may inhibit corrosion of iron pipes by forming protective coatings.

SODIUM (Na)

Description and Sources
Sodium is an abundant and widespread constituent of many soils and rocks and is the principal cation in many natural waters associated with argillaceous sediments, marine shales and evaporites, and in seawater. Sodium salts are very soluble and once in solution tend to stay in solution.

Sodium concentrations in natural waters vary from less than 1 mg/L in stream runoff from areas of high rainfall to more than 100,000 mg/L in groundwaters and surface waters associated with halite deposits in arid areas. In addition to natural sources of sodium, wastewater, industrial effluents, oilfield brines and de-icing salts may contribute sodium to surface and groundwaters.

Effect on Drinking Water

Sodium in drinking water may be harmful to persons suffering from cardiac, renal and circulatory diseases and to women with toxemias of pregnancy. Large sodium concentrations are toxic to most plants.

SPECIFIC CONDUCTANCE

Description and Sources

Specific conductance is a measure of the ability of water to transmit an electrical current and depends on the concentrations of ionized constituents dissolved in the water. Many natural waters in contact only with granite, well-leached soil, or other sparingly soluble material have a low conductance.

Effect on Drinking Water

The specific conductance is an indication of the degree of mineralization of a water and may be used to estimate the concentration of dissolved solids in the water.

SULFATE (SO_4)

Description and Sources

Sulfur is a minor constituent of Earth's crust but is widely distributed as metallic sulfides in igneous and sedimentary rocks. Weathering of metallic sulfides such as pyrite by oxygenated water yields sulfate ions to the water. Sulfate is also dissolved from soils and evaporite sediments containing gypsum or anhydrite. The sulfate concentration in natural fresh waters may range from zero to several thousand mg/L. Drainage from mines may add sulfate to waters by virtue of pyrite oxidation.

Effect on Drinking Water
Sulfate in drinking water may impart a bitter taste and act as a laxative on unacclimated users. According to the EPA National Secondary Drinking Water Regulations the maximum contaminant level of sulfate for public water systems is 250 mg/L.

ZINC (Zn)

Description and Sources
Generally, in streams and rivers, zinc is concentrated in sediments, but concentrations are quite low in running filtered water. In areas of soft, acidic water, however, pickup in the distribution system has been noted when comparing water samples from the treatment plant with samples at consumer's taps. This may result from the water flowing through galvanized iron pipes. In rocks, zinc is most commonly present in the form of the sulfide sphalerite, which is the most important zinc ore. Zinc may replace iron or magnesium in certain minerals and it may be present in carbonate sediments. In the weathering process, soluble compounds of zinc are formed and the presence of at least traces of zinc in water is common.

Effect on Drinking Water
Zinc is considered an essential trace element in human and animal nutrition. The recommended daily dietary allowances for zinc are as follows: adults, 15 milligrams each day, growing children over a year old, 10 milligrams each day, and additional supplements during pregnancy and breast feeding (check with your doctor). As far as drinking water is concerned, the EPA secondary drinking water

standard of 5 mg/L, assuming the common intake of two liters of water each day, would result in an intake of 10 milligrams of zinc each day, which is less than the estimated adult dietary requirement for zinc.

Concentrations of 40 milligrams in a liter of water gives the water a strong metallic taste (the technical description is astringent).

About the Authors

Dr. James M. Symons, Retired, is the Cullen Distinguished Professor Emeritus of Civil Engineering University of Houston, and holds a doctor of science degree from Massachusetts Institute of Technology. He is a diplomate of the American Academy of Environmental Engineers and a member of the National Academy of Engineering and American Water Works Association (AWWA). In addition to his academic career, he has worked for the US Environmental Protection Agency and US Public Health Service and has received numerous research and publications awards and honors. He is also the editor emeritus of *The Water Dictionary, A Comprehensive Reference of Water Terminology*, published by AWWA.

Nancy E. McTigue has worked actively in the water field since earning her bachelor's degree at Wellesley College and her Master's degree in environmental engineering from Stanford University. Her career has included positions at the USEPA and AWWA Research Foundation (now the Water Research Foundation). For the past several years, she has served as the director of research for Cornwell Engineering Group, an environmental consulting firm that provides services to water utilities worldwide. Her work has included technical reports, peer-reviewed articles, and presentations on various aspects of drinking water treatment optimization and management issues facing the industry. She has been an active and committed volunteer for AWWA throughout her career, serving on and chairing many committees and advisory boards. She has served as Trustee of both the AWWA Standards Council and the Technical and Education Council and is an Honorary Member of AWWA. She is also the editor of second edition of *The Water Dictionary, A Comprehensive Reference of Water Terminology*, published by AWWA.